The Fate of the Species

The Fate of the Species

Why the Human Race May Cause Its Own Extinction and How We Can Stop It

FRED GUTERL

BLOOMSBURY

New York Berlin London Sydney

Published by Bloomsbury USA, New York

All papers used by Bloomsbury USA are natural, recyclable products made from wood grown in well-managed forests. The manufacturing processes conform to the environmental regulations of the country of origin.

LIBRARY OF CONGRESS CATALOGING-IN-PUBLICATION DATA HAS BEEN APPLIED FOR.

ISBN: 978-1-60819-258-8

First U.S. edition 2012

1 3 5 7 9 10 8 6 4 2

Typeset by Westchester Book Group
Printed in the U.S.A. by Quad/Graphics, Fairfield, Pennsylvania

For Jude

Contents

	Introduction	1
1.	Superviruses	7
2.	Extinction	28
3.	Climate Change	50
4.	Ecosystems	84
5.	Synthetic Biology	95
6.	Machines	126
	Ingenuity	167
	Acknowledgments	187
	Notes	189
	Index	199

Introduction

A farmhand goes from pen to pen, dumping buckets of smelly gray slop into troughs. The pigs squeal with delight and tuck into their meal. Chickens, with free run of the farm, crowd in between the pink bodies to get their share of the grub. Clamshells and other detritus from the previous day's feeding lie scattered on the ground. In the middle of this chaotic scene, a boy squats over a basin of water washing cabbage—not for the animals, but for farmhands, who are hungry for lunch.

This was just one of many vivid descriptions that my colleague Sarah Schafer reported on a trip to China's Guandong province back in the fall of 2002. She was on assignment for *Newsweek International*, collaborating on an article on the then-recent outbreak of SARS, the deadly cold virus that briefly caused a scare when it spread quickly beyond China. The farm she visited was typical of those that had been popping up all over the province at the time to meet the growing demand for meat from an expanding Chinese middle class. The farm was a living petri dish for new viruses, bringing together birds, pigs, and humans in close proximity—a perfect breeding ground.

Over the next few years, the bigger story turned out not to be SARS, which trailed off quickly, but avian influenza, or bird flu. It had been making the rounds among birds in Southeast Asia for years. An outbreak in 1997 in Hong Kong and another in 2003 each called for the culling of thousands of birds and put virologists and health workers into a tizzy. Although the virus wasn't much of a threat to humans, scientists fretted over the possibility of a horrifying pandemic. Relatively few people caught the virus, but more than half of them died. What would happen if this bird flu virus made the jump to humans? What if it mutated in a way that allowed it to spread from one person to another, through tiny droplets of saliva in the air? One bad spin of the genetic roulette wheel and a deadly new human pathogen would spread across the globe in a matter of days. With a kill rate of 60 percent, such a pandemic would be devastating, to say the least.

Scientists were worried, all right, but the object of their worry was somewhat theoretical. Nobody knew for certain if such a supervirus was even possible. To cause that kind of damage to the human population, a flu virus has to combine two traits: lethality and transmissibility. The more optimistically minded scientists argued that one trait precluded the other, that if the bird flu acquired the ability to spread like wildfire, it would lose its ability to kill with terrifying efficiency. The virus would spread, cause some fever and sniffles, and take its place among the pantheon of ordinary flu viruses that come and go each season.

The optimists, we found out last fall, were wrong. Two groups of scientists working independently managed to create bird flu viruses in the lab that had that killer combination of lethality and transmissibility among humans. They did it for the best reasons, of course—to find vaccines and medicines to treat a pandemic should one occur, and more generally to understand how influenza viruses work. If we're lucky, the scientists will get there before nature manages to come up with the virus herself, or before someone steals the genetic blueprints and turns this knowledge against us.

Influenza is a natural killer, but we have made it our own. We have created the conditions for new viruses to flourish—among pigs in factory farms and live animal markets and a connected world of international trade and travel—and we've gone so far as to fabricate the virus ourselves. Flu is an excellent example of how we have, through our technologies and our dominant presence on the planet, begun to multiply the risks to our own survival.

The world lived for half a century with the specter of nuclear war and its potentially devastating consequences—radioactive clouds drifting thousands of miles and dust rising in mushroom clouds to the upper atmosphere, blocking sunlight and dispatching the human race in short order, just as the dinosaurs perished 65 million years ago after an asteroid impact. Then the Cold War ended. The Armageddon scenario has become less potent with each passing year. Nuclear weapons remain troublesome and dangerous. The missiles are still there, but the scenario seems remote.

Yet if anything the existential dangers have only multiplied since the end of the Cold War. Today the technologies we fear are not so much military as commercial. They come from biology and the information sciences, and they are behind our prodigious productivity. They are far more seductive than nuclear weapons, and more difficult to extricate ourselves from. The technologies we worry about today are the ones that form the basis of our global civilization and are essential to our survival.

The success of *Homo sapiens* has created new and terrifying risks that didn't exist a few decades ago. By our dominating presence on the planet, we are changing its geochemistry and its biology. We are upsetting climate systems—not just global average climate, but also an intricate network of regional weather systems—in ways we don't fully understand. Ocean current cycles, monsoons, glaciers, and rain forests could each turn suddenly, or in tandem. This may already have been set in motion.

As humans grow in number, we have also come into closer contact with other species, opening ourselves to new reservoirs of disease, and we've given these microbes a vast interconnected world in which to circulate. The conditions are ripe for a plague of twenty-first-century proportions, the Black Death that wiped out a third of Europe writ large and immediate. To fight these pathogens, we are racing to uncover mysteries at the chemical heart of biology, as we should, but at the same time we're opening new possibilities for mischief. We are beginning to create artificial life with technologies of molecular manipulation that will only become cheaper, faster, and more widespread, putting unprecedented power to destroy into the hands of individuals. Think of what a Unabomber-type lunatic could do now with a Ph.D. in microbiology and a few thousand dollars worth of lab equipment.

While the battle of the bugs rages, our own machines have become too complex for us to fully understand. A few decades ago the Internet was a novel technology shared among a small community of academics who knew and trusted one another. Its founders, taking such bonhomie for granted, built a house with no locks on the doors or windows, and computer scientists and engineers have been playing catch up with security ever since. Now, the Internet is the backbone of our global economy. Lives, not just livelihoods, depend on it. A new generation of software with the power to think and act much like people has thrown the future of this network in doubt, but our dependence on it only grows.

Scientists and engineers are generally upbeat people who love what they do and love to talk about it. But few of them, I've found, want to publicly discuss the darker implications of their work. Even those who are vocal in advocating policies to avert climate change or prepare for a bioweapons attack prefer to dwell on what can be done to prevent bad things from happening, rather than what those bad things might be. This is not universally true, of course. I once asked Stephen Pacala, a professor at Princeton, what the consequences might be if the permafrost of Canada and Alaska and Siberia were to melt and release the methane and carbon they now keep locked in the ground. His answer: "All sorts of monsters would come out."

This book is about those monsters.

The most terrifying monsters are the ones that we have created with our technology and our dominance of the planet. Killer volcanoes or asteroid impacts or the sun going supernova are truly terrifying, and they're all remotely possible (perhaps not so remote, when it comes to asteroids). This is mainly a book about technology, about artifice, about the unintended consequences of the success of *Homo sapiens*. What I'm aiming to do is tell some stories about real dangers we face. I won't give you a balanced view. I will intentionally ignore the bright side of these issues and focus on the question, How bad can it be?

I have no interest in making predictions. Because the world is unpredictable, it's a good exercise to consider the worst case. It forces us to think about where we fit into the world and ask where we stand in relation to the history of life on Earth. Here we will explore the products of our own success, and things that we can do something about, if we care to. These scenarios are all the more interesting and scary because they could turn out to be true.

When my father was born in 1924, the world population stood at about 2 billion people. By the time I was born, the population had increased 50 percent to about 3 billion. When my daughter was born, the population was about 5.5 billion. A population cannot keep growing faster and faster forever. (Neither can property values, unfortunately.) The truism is easier to accept for bacteria or rodents or sea gulls than for humans, but that doesn't make it any less true.

Doomsayers have a way of coming and going. Thomas Malthus got it wrong in the eighteenth century and Paul Ehrlich in the 1960s. We humans have a history of bucking the system. Perhaps we can engineer the planet and our own population so that we avoid a crash. We're going to talk here, however, about how a crash could happen, what form it might take, and what could trigger it.

Optimism, I've come to believe, is an outlook, a state of mind that is partly reason and emotion, partly a product of personality. I tend toward the techno-optimistic side of the spectrum. I also think optimism is our best weapon. There's no going back on our reliance on computers and high-tech medicine, agriculture, power generation, and so forth without causing vast human suffering—unless you want to contemplate reducing the world population by many billions of people. We have climbed out on a technological limb, and turning back is a disturbing option. We are dependent on our technology, yet our technology now presents the seeds of our destruction. It's a dilemma. I don't pretend to have a way out. We should start by being aware of the problem.

This book starts with a story of ordinary flu and the pandemic of 2009—a mild outbreak that, had a few bits of genetic code mutated a bit differently, could have been a global catastrophe. Then we step back and look at the question of whether we may be in the beginnings of a "mass extinction event"—a grand die-off of species. If we are, and if there is no way to stop it, it will spell the end of the human race. After that, we look at climate, and at the possibility that things could change drastically and suddenly for the worse—more suddenly than most scientists are willing to talk about.

Climate change has put the planet in a precarious position, balanced on the edge—what could push it over? That is the subject of the next few chapters. The food web that sustains our system of agriculture is out of whack, making our food supply vulnerable to disease and disruption. Our biotechnological prowess has made us vulnerable to horrific designer diseases. And we have built our global economy on a technology—the Internet—that is unreliable and ridiculously vulnerable. And now we are developing machines that can mimic the thoughts and deeds of humans, creating the possibility that they could turn on us, or be turned against us. Finally, we'll talk about what must be done to neutralize these threats. We cannot use the situation we've inadvertently created to turn away from technology. Rather, we must redouble our efforts to use it well.

As I was writing these words, in October 2011 in New Jersey, snow was falling outside my window. All told we got about seven inches. But the funny thing was, the leaves were still on the trees. The fall had come late that year, and the snow had started early. It made for a beautiful sight, the snow on the green and yellow and red leaves.

But the snow hadn't fallen for more than a couple of hours when the whole thing took a sinister turn. The snow was heavy, and the leaves gave it plenty of surfaces to cling to. As a result, branches started to fall. My wife and I went to the window to look at the snow fall and in just a few minutes, three or four big branches in our backyard give way under the weight.

I grabbed my camera and went for a walk around the block. I took care not to walk under any tree, because every few seconds you could hear the crack and whoosh of falling branches. I walked two or three blocks, and saw three power lines that had come down from fallen trees.

We were lucky. We only lost power for about twelve hours. Some of our neighbors were without power for seven days. The local power company called in reinforcements from as far away as Georgia to help restore downed lines. The governor declared the state a disaster area. Supermarkets were shuttered. Trains didn't run for several days.

All this time, I couldn't help but think of how a small change in timing—the early snow and the late fall—could turn a minor weather event into a crisis. This one fortunately passed: Power was restored, the snow eventually melted, life returned to normal.

And I went back to work, exploring the valley of dry bones.

CHAPTER 1

Superviruses

YOSHI KAWAOKA SEEMED DESTINED for mediocrity. At the end of each academic year when his high school class would sit for their exams, Kawaoka never managed to snare the top grades. In his junior year, he placed near 375th out of 450 on a major exam. In Japan's rigidly hierarchical educational system, he had no hope of making it to Tokyo University, which skims off the top 1 or 2 percent. His middling scores earned him a place at modest Hokkaido University, in the north. He bought himself a parka.

At Hokkaido, Kawaoka was an indifferent student. He studied veterinary medicine and bacteriology, but he couldn't get too excited about it. One of the professors at the graduate school, Hiroshi Kida, took an interest in the young researcher. There were smarter students, and more ambitious students, but Kawaoka had the personality for lab work. In the world of virology, where you're dealing with tiny pathogens too small for the eye to see, and about which little is known, often the only way to move forward is to run dozens and dozens of experiments, keeping tabs on them all, tabulating and tracking all the data over weeks and months, without losing sight of the often small piece of knowledge you hoped to extract. After all the tests and the tabulating, the results are often murky, or something goes awry—someone forgets to wash a beaker, and a minute contaminant seeps into the experiment—and you have to repeat the whole thing, exactly as you did it the first time.

The smartest students tended to grow bored with such detail and repetition. They would work for a while in the lab and opt for other fields with a quicker payoff, such as medicine or teaching. Kawaoka, though, seemed to respond to the rigors of the lab bench. Almost uniquely among Kida's students, he would enter the labyrinth of experimental data and emerge from the other side with some clarifying result. Still, when it came time to graduate, Kawaoka was toying with the idea of leaving the field and following his father into business.

Perhaps partly because Kawaoka was a favorite student, and partly to keep him interested, Kida recommended him for a plum postdoctoral research position in an influenza research lab in Tennessee. Dr. Robert Webster ran a small team of influenza researchers at St. Jude Children's Research Hospital in downtown Memphis. Doing research on a commonplace bug seemed a bit dull to Kawaoka, but he had always wanted to travel to the United States, and it would be a good opportunity to practice his English. In August 1983, he boarded a flight to Memphis and curled up with a copy of a book on Ebola, the killer virus from Africa.

Kawaoka didn't read English well at the time, but he plowed through the accounts of disease and gruesome death in his characteristic dogged style. Ebola is one of the deadliest human pathogens ever known—or at least the most grisly. Like any virus, it works by penetrating the membrane of a human cell and hijacking its biochemical machinery, turning the cell into a tiny Ebola virus factory. Whereas most human viruses tend to infect one part of the body—influenza, for instance, usually zooms in on the lungs and upper respiratory tract—Ebola attacks just about every type of cell, reproducing with no apparent sense of limit. No matter where the infection starts—in the blood, liver, lungs—the Ebola virus replicates with such unstoppable efficiency that they physically overwhelm each and every cell they infect. Through an electron microscope, an Ebola-infected cell looks like a big ball of spaghetti, each strand a virus "particle" straining to find a new host.

Ebola is so virulent that an infection begins to have a structural effect on the body. Membranes dissolve. Organs turn to mush. Blood oozes from the pores of the skin. Some Ebola strains kill 90 percent of victims. The gruesome results—massive internal bleeding, with every membrane and organ turning into a gelatinous goo—gave Kawaoka a big motivation for parsing one English sentence after another.

Kawaoka arrived at the Memphis airport bleary-eyed from reading. He could hardly make out what customs and immigration officials were trying to say in their Southern drawls. He must have seemed like an odd character to them. In clothing, his taste runs to shades of gray—charcoal trousers, black shirt, jacket, and shoes. With a slight frame and closely cropped goatee, he looks like a Japanese Johnny Depp.

At the lab, Webster assigned Kawaoka to gather data for a pet theory: that human influenza viruses come from waterfowl. For most of the time, a virus lives as a harmless stowaway in the cells of ducks. Once in a while, the virus will mutate in such a way as to jump to another species—often a pig, which both bird and human varieties of influenza seem to infect with ease. Here's how it might go: A flock of influenza-infected (but outwardly healthy) ducks will alight for a drink from the trough of a livestock farm in China, passing the virus along to the pigs who are drinking the same water. The pigs will in turn pass the virus (through their feces) to the chickens. A farmhand comes along and sneezes on the pigs, passing along a human influenza virus. Now the poor animals have both bird and human strains in their cells, where the viruses swap genes like boys trading baseballs cards. This "genetic reshuffling" is characteristic of the influenza virus; it happens ceaselessly. As China's middle class has grown bigger and richer, it has demanded more meat, and to keep up, farms have grown more like factories in scale. The virus's spawning grounds have grown as well. The bigger the farm, the more genetic opportunity influenza has to mix and match and mutate into a form that can do some damage in the human population.

Webster found the notion that influenza had a vast safe harbor in wild birds unsettling. It meant, for one thing, that the disease could never be eradicated from the human population, because it would always be replenished with an endless supply of new viruses from birds. It suggested a frightening and even sinister threat—the kind of thing that Joshua Lederberg, the late Nobel Prize–winning biologist, had in mind when he wrote that "the single biggest threat to man's continuous dominance on the planet is the virus."

At the time, however, this was all just Webster's hunch. Kawaoka's job was to investigate. He ran lab tests on blood and tissue samples taken from wild ducks. First he would add an enzyme that broke open all the cells in the sample, spilling their genetic material. Then he would separate out the influenza viruses and put tiny droplets containing flu bugs into cultures, where they would replicate. Finally, he would run tests to determine what strains the ducks were carrying. The goal was to make a kind of map that showed how the viruses jumped from one wild duck to another, and eventually to other species.

Barely two months after Kawaoka arrived, Webster announced abruptly one day in November that he was getting on a plane for Pennsylvania. An outbreak of bird flu on some chicken farms was killing the birds by the thousands. This genuine emergency had potentially big commercial consequences—poultry farms from Pennsylvania to Virginia were at risk. It would also give Webster a chance to see a real avian influenza outbreak in action, an opportunity he rarely missed. The immediate worry was that people would die by catching a fatal bird flu from close contact with the chickens. At the epicenter of the outbreak was a farm run by a family of Mennonites in Lancaster, outside Philadelphia. Children on a family farm are often the jumping-off point for bird flu. They tend to treat the chickens as pets, cuddling them and kissing them, giving the influenza virus ample opportunity to make the jump.

A somewhat more abstract concern had also begun to fester in the back of Webster's mind: that this virus, so catastrophically fatal to birds, could make its way into a human host—one of the children on the farm perhaps?—and once there, genetically shuffle itself into some kind of new human pathogen with unprecedented power to kill. These are the kinds of dark places Webster's research was leading him to.

Webster wanted Kawaoka to be ready to start studying the new virus. "I'll FedEx tissue samples," Webster said. Then he picked up his bag and headed for the airport.

A CHICKEN INSPECTOR with a modicum of experience can tell as soon as he walks into a poultry house if something is amiss. Normally, there's a reassuring murmur of clucking and the fluttering of feathers. Sick chickens make a different sound—like thousands of tiny people sneezing and coughing.

When Gerald Fichtner, the Northeast regional manager for the U.S. Department of Agriculture, walked into the Mennonite farm in Pennsylvania, he heard something even worse: silence.

Fichtner had just driven all the way down from his home in Schenectady, New York, with the bag his wife had packed six months earlier. Since April, he had been aware of a bird flu virus making the rounds of poultry farms in Pennsylvania. But the virus was deemed to be "low pathogenic," meaning it gave the birds the sniffles and a cough, but didn't kill them. As long as it had mild effects, it fell under the jurisdiction of the State of Pennsylvania, but Fichtner knew from experience that any flu virus was potentially troublesome. When the call came in early November, he had been expecting it. He had told his wife and five children that he wouldn't be home again for months.

The desperate Pennsylvania chicken farmers told Fichtner that the mortality rate of this new flu strain was 100 percent—it killed every last chicken that got infected.

The virus was probably moving rapidly through the poultry population, spreading in an arc from Pennsylvania to Maryland and down to Virginia, the heart of the poultry industry. If Fichtner had had a vaccine ready for this strain, he might have been able to vaccinate chickens in surrounding farms to keep the disease from spreading. But none had been prepared ahead of time, and the Department of Agriculture needed six months to get one ready.

His only choice was "depopulation."

That presented a particularly grisly logistical problem. How exactly do you go about eradicating thousands of chickens at a farm in a way that doesn't spread the disease? You have to kill the animals antiseptically, so nothing infected is left around. And you can't bury the carcasses, because animals might dig up the remains, exposing the virus to other birds. The chickens had to be incinerated, but the farms didn't have furnaces. He had to find a way of killing the chickens on the farms and transporting the corpses to county incinerators.

Fichtner's nasty little stroke of genius was to use a combination of Dumpsters and carbon dioxide. His crew would park an industrial Dumpster near a chicken house, right underneath a second-story doorway. They'd line the Dumpsters with twenty-foot rolls of plastic and open the valve of a canister of carbon dioxide, allowing the gas, which is heavier than air, to settle to the bottom. Then forty or fifty local Amish teenagers—dressed in sanitary overalls, masks, gloves, and hairnets—would run through the chicken house gathering up the birds four to a hand and flinging them from the second-floor door into the Dumpster. The chickens would stagger around for a few minutes, clucking and sneezing, then collapse from the lack of oxygen. Fichtner typically got fifty thousand chickens to a dumpster.

Webster arrived in the middle of this chicken Armageddon. He dispensed wisdom about disinfecting and keeping workers from exposure to the virus. The "bunny suits" were Webster's idea. He also advised Fichtner that whenever one of the trucks taking the chickens to the incinerators hit a pothole, he had to make sure no carcasses fell off onto the road. If any did, they had to stop, don the bunny suits once again, pick up the birds, and disinfect the street. Webster didn't want any unsuspecting sparrow picking up the virus and spreading it through the wild-bird population, or he'd have a different kind of catastrophe on his hands.

Kawaoka, already bored with Memphis, signed for the FedEx packages with a sense of relief. He took them down into the basement to St. Jude's biocontainment lab and through the heavy, padlocked door to the air lock. He discarded his clothes and donned a jumpsuit with a breathing apparatus, which would minimize the possibility of accidentally breathing in a droplet of air that contained the virus. He passed through another padlocked door. Once inside, he opened the boxes and got to work.

In the basement lab, Kawaoka had a wall of cages, each containing a healthy chicken. At the lab bench, he extracted the virus from the samples, grew cultures, and used them to infect each of the chickens. Then he waited for signs of flu.

Each day he came back to check up on his birds. As expected, they began to cough and wheeze. A few days later they grew listless.

After a week or so, every last bird was dead. The farmers hadn't exaggerated: the mortality rate was 100 percent. That was the first surprise.

The next step was to cut open one of the dead birds and see where the virus liked to live. In mammals, influenza is a disease of the respiratory tract; in birds, it typically infects the gastrointestinal tract. But precisely where the virus tends to settle is a matter of great interest to virologists because that can give valuable clues as to the virus's behavior, transmissibility, and so forth. Kawaoka took a knife and made an incision from just under the chicken's neck and down through its belly. He had performed this procedure hundreds of times on many different birds, but what he saw this time was not typical.

The blade went in, and out oozed a blackish, gelatinous goo. The virus had consumed the bird's organs.

"Just like Ebola," he thought.

THE STATE OF medical technology, when it comes to flu, has not moved much in the past century. To be sure, virologists have done miracles in the lab—they've peered into the very genetic machinery of influenza, they've invented their own strains—but they don't understand how the virus works.

More to the point, they have no way of reacting quickly enough to stop an outbreak that takes them by surprise. Despite all the miracles of recent biology, flu vaccines are still largely made the old-fashioned way—by growing them in chicken eggs. If doctors don't get months' lead time for the next flu bug, they might not have time to prepare a vaccine. At the moment, there is no reliable way to speed that process along.

For this reason, a flu virus doesn't need to kill 100 percent, or even 90 percent, of its victims to be devastating. A new influenza virus with a 60-percent mortality rate—a rate that some virologists contend is entirely possible and would be consistent with the known rate of mortality from bird flu—could, if it struck without warning, conceivably result in hundreds of millions of deaths. And the world's doctors would be powerless to stop it.

We are, in a sense, Fichtner's chickens, waiting for the fall.

What started as a niggle at the back of Webster's mind in the early 1980s has in the intervening twenty years taken on the solidity of conventional wisdom.[1] Webster's theories of human influenza viruses deriving from bird varieties has been amply borne out in the research. We are surrounded by influenza viruses that live in birds, and this reservoir poses a constant existential threat to humans. The prospect of eradicating this threat is essentially zero. From the history of bird flu in the last few decades, you might get the impression that influenza has a demonic intention to wreak havoc. The virus seems to be straining to evolve into a new form that targets the world's human population.[2]

Of course, influenza viruses do not have intentions, demonic or otherwise. They are not even really alive. They are incomplete passages of genetic material—RNA, in the case of flu, a molecule similar to the more widely known DNA. They cannot perform the most fundamental task of reproduction without help. Influenza is a parasite of sorts that relies on full-fledged cells of its host to complete it, and on a rapid evolution that gives it the ability to mutate, quickly changing in response to new circumstances. People who study viruses in general, and flu in particular, sense that the world has changed in the past few decades, and is continuing to change, in ways that have tilted the playing field to influenza's advantage. The growth of livestock farms in Asia, the globalization of trade, the rise of the middle class in such places as China, India, and Mexico, the shift in climate—all of these factors may have created a novel habitat for influenza. That is the theory, anyway. What we know is that evolution is opportunistic, and that with a new niche eventually comes a new disease.

The question of whether a niche is open for a global killer of unprecedented proportions is not settled by any means. Many virologists remain to be convinced that a highly lethal flu virus would survive long enough in "the wild" to spread as a pandemic. If a pathogen kills its host too quickly and efficiently, it eventually burns itself out. The history of influenza, the argument goes, bears this out: most cases of flu are mild, and casualties are concentrated in the very old and the very young. We are not Fichtner's chickens precisely because we don't live in a chicken coop. On the other hand, the history of bird-flu outbreaks gives many scientists pause. A few close calls have convinced Kawaoka, Webster, and others that bird flu could turn into a lethal human pathogen.

The first episode started in May 1997 when a three-year-old boy from Hong Kong came down with a fever, a cough, and a sore throat. His physician thought it was a routine illness, but when his symptoms worsened, his parents took the boy to a local hospital. The staff were flummoxed—the illness fit no profile they were familiar with—and referred him to Queen Elizabeth Hospital in Kowloon, where he was quickly moved to intensive care. Twelve days after showing symptoms, the boy died of multiple organ failure.[3]

The boy's blood sample, meanwhile, took on a life of its own. Queen doctors referred it to a virologist in Indonesia, who passed it to a colleague in Holland, who, after more testing, sent it along to Eric Claas, a protégé of Robert Webster's. Claas, one of a handful of flu specialists equipped to test for every known species, came back with a startling result: the boy had contracted H5N1, a type of flu hitherto known only in birds.

By then, the boy had been dead for three months, but the scientists had lost the link between patient and sample. Scientists at the CDC heard about the case and confirmed it. A team flew out to Hong Kong. Their first task was to see if the boy had gotten the virus from chickens or from another person. It was dumb luck that the virus hadn't acquired the ability to jump from human to human; if it had, it might have been around the world and back by the time Claas had identified it. All told, eighteen people caught the H5N1 that year, half of them from contact with birds. Six people died.[4]

The next day, Webster got on a plane for Hong Kong. He met up with longtime friend Ken Shortridge, who held an appointment at Hong Kong University, and the two went around to the poultry markets in Hong Kong, where farmers set up their booths in the wee hours of the morning. They were looking for birds with signs of respiratory infections, and when they found one, they'd swab the mucus, wipe the swab onto a culture dish, label it, and move on. Then they'd go back to Shortridge's lab and run tests, and for the ones they couldn't analyze, they'd send them back to St. Jude's. Gradually they amassed a catalog of viruses, and they began to see that the bird flu virus that killed the poor boy had been circulating among livestock farms for years. Webster continued to make trips, sometimes sending Kawaoka in his place, and sometimes a young Chinese native named Guan Yi.

At the time of the 1997 outbreak, Guan was working in Webster's Memphis lab. Webster quickly realized that Guan would be invaluable in establishing a flu surveillance program on the Chinese mainland. In 2000, Guan moved back to Hong Kong. He began making regular visits to the poultry markets in Hong Kong and Guangdong, and tracking the progress of viruses.

It didn't take Guan Yi long to figure out what needed to be done about the rise of bird flu cases. Guan had grown up in mainland China's Guangdong Province, among the livestock farms and the animal markets of Guangzhou. He knew well how much the Chinese prize their chicken, duck, and pork (not to mention civets and dogs). And he knew well the southern-Chinese tradition of live meat. In the live-animal market of Guangzhou, cages are often stacked one on top of the other, each with its own bird. These birds live cheek by jowl with one another, and they breathe in droplets of saliva when nearby birds sneeze, some of them laden with influenza viruses. The birds swap viruses among them.

Before the birds arrive in their cages, they're kept on small farms, where they roam freely among the pigs. Influenza viruses get swapped back and forth between the birds and the pigs—not quite as easily as among birds alone, but there's so much contact that it happens quite often. Pigs run around on the farms and defecate on the ground, where the chickens feed, and the chickens defecate where the pigs like to roll around in the mud. Then afterward they both knock off to the trough for a nice long drink. Walk into one of these farms and you are impressed by two things: the chaos, and the stench. All this biological intimacy means that the microbes are intimate as well. Livestock farms in Guangdong Province may be the most concentrated of all populations of influenza. It's a giant gene-swapping convention. To stop the spread of the virus, mass slaughter of poultry was carried out.

Guan's vigilance, however, wasn't enough to forestall another outbreak of bird flu in late 2003. Hundreds of thousands of birds had to be culled throughout Southeast Asia. In 2004, at least 44 people caught bird flu, and 32 died. All told, between the end of 2003 and the summer of 2005, 112 cases of bird flu were reported to the WHO; 57 people died. Worryingly, the outbreak included the first cases of human-to-human transmission, among family members.[5]

An influenza virus particle works much like a hijacker—by breaking and entering a human lung cell and taking over. It's a delicate and complex operation. The virus first has to enter the lungs on a current of air. It has to penetrate the membrane of a lung cell and inject its genetic material. It has to insert itself into the cell's genetic machinery, altering the cell's biochemical clockwork so that instead of performing all the day-to-day tasks of staying alive, it starts manufacturing copies of the virus. When that job is done, the virus particles collect in the lungs, ready to go out like shock troops when the victim sneezes and find another victim, and the cycle starts anew.

At each step, the virus needs to crack some kind of code—it needs to present the proper sequence of nucleotides to get it past the next line of defense. One sequence will give the virus just the right properties to float in the air without succumbing to the stress of its environment, long enough to find a victim. Another sequence will give it the ability to latch on to a lung cell and pierce the membrane. Another sequence gives it the ability to insert itself among the host cell's genetic material and divert it. But it's probably more complex than that, with many sequences playing a role in particular traits.

Influenza is the master of change—its survival as a species hinges on its ability to alter its genetic makeup. Each nucleotide of the influenza genome is subject to mutations—sometimes these occur randomly, sometimes they involve swapping whole genes with other viruses. This is true of nucleotides in all organisms, of course. But flu viruses are particularly swift at this evolutionary roulette, which gives the virus an alarming ability to target itself as a human pathogen. It is influenza's way of probing for a new pathogenic niche. It's like playing a slot machine. Each season we pull on the handle and wait to see if we've won a jackpot. Usually we get mixed fruit, and a mild outbreak. Once in a while, we get a small jackpot (a mild pandemic, say, such as the 2009 outbreak). The slot machine could also come up all bananas—some combination of nucleotides that turn a mild human pathogen into a killer on a scale that nobody has known before. The virus, by recombining with other viruses, swapping genes, and taking on random mutations in its genetic code, keeps trying to find this jackpot combination—a bug that spreads easily and quickly, but strikes with deadly virulence. It's possible that the genetic slot machine hasn't yet produced its worst—and that the deadliest human influenza virus awaits us.

Kawaoka got an inkling of how dangerous this evolutionary slot machine can be back in the 1980s. When the dust settled on the Pennsylvania outbreak, he turned to figuring out how a bird flu virus that was mild in April 1983 turned into a killer by November. He decided to compare the two viruses' genomes. A genome is the complete genetic alphabet of an organism—a sequence of individual letters, or nucleotides, that make a particular strain of influenza what it is. In the case of flu, the typical virus has about fourteen thousand nucleotides. Included on that stretch are eight genes and some random passages of unknown purpose, some of which govern the creation of proteins at some stage in the virus's life. The difference between the April 1983 virus and the November one—between a nuisance and a death sentence—came down to a single protein.

"What this tells you," says Kawaoka, "is the highly pathogenic virus was generated from a single mutation. And it tells you there are many sources of highly pathogenic influenza viruses. It's all out there in birds."

IT MAY SOUND like science fiction, but isn't. We have lived through the scare of a potentially catastrophic influenza outbreak: the pandemic of 2009.

Most people probably remember the outbreak as an overreaction by health officials. Sure, there was an initial scare in April 2009 when new cases started coming in and scientists were talking about a new strain of flu that could be dangerous. By the time the World Health Organization declared the outbreak a pandemic in the summer, everybody knew it was a mild form—even milder than the typical seasonal flu.[6] In Europe, an official accused the WHO of playing into the business plans of the big pharmaceutical companies, who supposedly stood to make big bucks from vaccines and antiviral medications. In the end, many European and U.S. vaccine supplies stayed on the shelves, unused.[7]

But the experts didn't overreact to the 2009 outbreak. The virus came from nowhere, and by the time it made it onto the radar screen of health officials, it was already well on its way to spreading far and wide. "H1N1 caught us all with our pants down," says Webster. "Not one virologist had the slightest suspicion." A year before the outbreak, he says, "no one, but no one, would have said that H1N1 would have been the next pandemic." In spring 2009, health officials must have felt as if they were staring into the abyss.

"This time we got damn lucky the virus was only mildly pathogenic." If H1N1 had truly killed as effectively as the 1918 flu, it would have been "total disaster," Webster says. "You wouldn't get the gasoline for your car, you wouldn't get the electricity for your power, you wouldn't get the medicines you need. Society as we know it would fall apart. There wouldn't be a hell of a lot scientists could do for you in the first wave."

Let's look at the 2009 outbreak again, through our doomsday lens. It will give us a contemporary look at what the beginning of the end might actually look like. Then we'll look at what might have happened if nature hadn't decided to let us off the hook.

ADELA MARIA GUTIERREZ of Oaxaca probably thought she had a bad cold. She worked as a pollster at a government office in Oaxaca, going door to door to gather census data. She lived with her husband and three children and had no health insurance, so when she began running a fever she went to a private doctor, who prescribed antibiotics. Five days later, she was coughing up blood. She returned to the doctor, who gave her something else, but the very next day her hands and feet started to turn blue. She went back to the doctor a third time, and he sent her to the hospital. The diagnosis was pneumonia. On the second day of her hospitalization, she developed hypoxic encephalopathy, which means her brain wasn't getting enough oxygen. Doctors put her on a ventilator. On April 13, eleven days after her first doctor's visit, she suffered cardiac arrest and died. In those last few days, it wasn't the illness they were fighting so much as Gutierrez herself, or more specifically, her immune system. It was working so hard to fight off the virus, it was killing her in the process.[8]

About the time Gutierrez died, Dr. Richard Besser ran into Anne Schuchat in the corridors of the Atlanta headquarters of the Centers for Disease Control, the quasi-governmental agency charged with overseeing the nation's health. Besser had taken over at CDC only a few months before. He had been head of the center's terrorism and emergency-preparedness division when the newly installed Obama White House staff had tagged him to run the agency. Tom Daschle, the former Senate majority leader, was Obama's favorite to become the next secretary of health and human services, but the appointment was dragging on.[9] Schuchat was the CDC's influenza point person—she ran a weekly influenza strategy meeting at CDC—and she was concerned. An unusual outbreak had shown up on her radar, and a few odd coincidences as well. "Ann said, 'I think you should come to the meeting on Wednesday. There are these two cases of swine flu from San Diego. There's going to be a presentation on that. It's something I really think you need to engage in.' So I went to the meeting."

By then, a few other cases had turned up in Texas. There had been an outbreak in Mexico, and reports of a higher-than-usual mortality rate had begun to circulate. "There was a concern that these might be related," said Besser. The CDC activated its emergency operations center.

You might think that the world's influenza surveillance network is highly formalized and automated, with information flowing from the remotest doctors' offices to command and control centers at the speed of light. But the reality is different. When it came time for the CDC—the nation's disease watchdogs—to find out what was going on, they drew on a loose connection of acquaintances—"back channels," as Besser put it. After the meeting, Besser's lieutenants got on the horn and called around. They called their colleagues in the public health labs in Winnipeg and Mexico City, compared notes, and concluded that they were all probably dealing with the same virus, and that it was something completely new.

Something new—words no influenza specialist wants to hear.

A ripple of fear began to spread through the public health community. Not only was this virus new, it also bore a striking resemblance to the deadliest flu virus on record—the H1N1 bug that caused the 1918 pandemic.

It was a frightening precedent. In 1918, the H1N1 virus burned through the world in less than two years, despite the lack of air travel, leaving 50 million to 100 million people dead in its wake. (Nobody knows precisely how many died. Information about influenza deaths was hard to come by back then, and in many ways it still is.) At the time, the world held 1.6 billion people. By simple extrapolation to the current population of more than 7 billion, a similar disaster today would leave 180 million to 375 million dead.

THURSDAY, APRIL 23, was take-your-daughters-and-sons-to-work day. New York City's public health office was crawling with kids. "My office was extremely chaotic," says Marci Layton, the assistant commissioner for communicable diseases. "There were kids everywhere." She and a colleague, with her kids in the room, took a conference call that day with the CDC, and that's where she first heard of the outbreak in Mexico and the oddball cases in San Diego and Texas.

That was also the afternoon a sharp nurse at St. Francis prep school in Queens called the health officials to alert them about a number of students who had come into her office complaining of symptoms much like strep throat. By the time Layton had heard of the school outbreak, school was over and the children had gone home. With cooperation from the principal and the school, they got phone numbers and called the children's parents at home.

The next day, Layton came to her office and read her e-mail. Canada's Winnipeg lab and the CDC had confirmed that the virus that had caused the outbreaks in Mexico and the United States was indeed swine flu—H1N1. Layton immediately dispatched health workers to St. Francis to get samples.

About five P.M. the samples arrived in the city's lab, where health workers were waiting. They worked into the night and by two A.M. had their answer: St. Francis was most likely having an outbreak of H1N1, the same as the virus in Mexico.

Health authorities in New York City and Atlanta then had to make some decisions—without knowing how virulent the virus in question was. Remember, virologists can't tell, even if they sequence a virus's genome, whether it's going to be a lion or a pussycat.

The quickest way to spread a flu virus is to keep the schools open. Should New York City close St. Francis? Should it close all the city's schools? What about schools across the nation?

The question was politically charged. With so little known about the virus, Besser's instincts were to err on the side of caution. "There was a school outbreak in New York. We were concerned about children being a way of spreading flu within the community, as well as children being likely victims of this flu. We worked on drafting guidelines for this pandemic. The initial guidance was that if you have a confirmed or suspected case, you would shut your school down until you had a chance to investigate this further. This guidance was developed and put on the Web." Besser's team then revised this recommendation: The CDC posted on its website advice to close schools for two weeks if flu was suspected.

About this time, Besser was called to Washington to brief the president.

Besser went to the White House and gave a short talk to the president and his cabinet and answered their questions. In the end, the White House persuaded Besser to reconsider the recommendation. Instead, the

CDC would wait a week to see what data emerged about the Mexico outbreak—such as what proportion of people who had contracted the virus died.

It was a political compromise to forestall panic, at the risk of giving the virus time to spread. It turned out to be the right call. The very next day, according to Besser, the CDC got information from Mexico that the mortality rate of the virus was lower than initially feared, and that closing schools wouldn't be necessary after all. St. Francis remained open. No harm was done.

We all know how the 2009 virus story ended. We lucked out. By June, data from Mexico and the United States showed that the new H1N1 virus was mild; early high mortality rates in Mexico had been exaggerated. When the virus resurfaced in the northern hemisphere in the fall, it remained mild—unlike the bird flu virus that returned with devastating effect to the poultry farms of Pennsylvania in 1983.

But what would have happened if the H1N1 virus of 2009 had been as virulent as the 1918 flu, or even the bird flu virus of Pennsylvania? Kawaoka's analysis of the 1983 bird flu showed that a single mutation had turned a mild disease into a deadly one.

Here's a scenario that could answer that question. Imagine that the data coming back from Mexico hadn't been such a relief. Here, based on what we know did happen during the 2009 pandemic, is how things might have gone.

By early May, schools in New York City and across the nation have been open for two weeks, as officials grew more and more concerned that the virus could be lethal. The kids at St. Francis who were part of the original outbreak have been moved to intensive care, and the first one dies—a healthy young soccer player who had never even logged a sick day. By the time the CDC calls for school closings, parents had already started keeping their kids home. They drop the kids at the mall and the rec centers, where they continue to spread the disease.

The information from Mexico confirms the worst fears of the epidemiologists: the new H1N1 flu is indeed deadlier than anything we've known since the 1918 flu, and possibly worse than that. Mortality rates of 60 percent—the same as for those few unfortunate souls who contracted H5N1 bird flu in 2004—are confirmed for H1N1. Reports from Mexico City wind up on television screens around the country. Mexican hospitals are overwhelmed with intakes, people are dying in the hallways, and supplies grow short. Politicians go on television to warn people not to panic, that hospitals in the United States will not run out of supplies or beds.

The workers at the New York City health department are mobilized. The city's labs on First Avenue are now staffed continuously, and infantry are posted outside to maintain order. The military and the FBI are also present inside the labs, because the crisis has provoked the obvious question: is this a natural outbreak or the work of terrorists who have fashioned a deadly bioweapon disguised as influenza? Meanwhile, city scientists are busily testing samples to track the virus's progress through the city, and to get a handle on any mutations that might make things better, or worse.

In Washington, the head of the CDC steps off a plane and makes his way to the White House to brief the president and his cabinet officials. He has already met with the heads of influenza and emergency preparedness. The president and his staff want to know how bad the outbreak could get, and what steps should be taken to head it off. The CDC head tells them the frightening truth: that millions of people will surely die, and nothing can stop it.

There is a silence around the table as the president takes stock of what the CDC head sitting across from him has said. "Should we close the borders?" the president asks.

ONE OF THE first things that happened during the 2009 outbreak is that politicians started calling for the nation to close its borders with Mexico. Fortunately cooler heads prevailed. This policy would have been misguided on at least two counts. For one thing, the 2009 H1N1 virus had already crossed the border by the time officials knew about it. It is also doubtful that the virus originated in Mexico—the more likely source was U.S. pig farms, where precursors of the virus rattled around for years before picking up the mutations they needed to take flight.

THE CDC HEAD answers the border-closing question without hesitating. "No, sir," he says. "Containing the virus is impossible at this point. By the time we had even heard of the virus, it was already too late. It will sweep through the entire population of the nation, and the world, in weeks."

Panicked politicians get the upper hand and close the nation's borders. Flights to the United States are canceled, ship traffic halted, and the highways between the United States and Mexico and Canada blocked. Although domestic air travel continues, at least for a while, American citizens traveling to Europe and Asia are stuck for the foreseeable future, and visitors to the United States similarly have to wait the crisis out.

"Can we vaccinate people?" asks the president.

"We have been working on a new technology for making quick vaccines, Mr. President, but it is finicky. It might work, but it might not. If it works, we could have vaccines in a month. If not, it will take six months. In either case, we will have casualties."

THE FIRST DOSES of 2009 H1N1 vaccine were not available until late October, and they weren't widely available until well into November. This was no screwup—vaccines take months to manufacture. First virologists have to isolate the virus they want to vaccinate against. Then they have to come up with a strain that will cause the immune system to kick into play—a strain that has, in other words, the same heamagglutinin and neuraminidase surface proteins (the *H* and *N* in H1N1) as the pathogenic strain. Then they have to neutralize the virus so that it doesn't cause illness. The traditional way of performing this combination—or *reassortment* of genes, in the scientific parlance—is to insert the two strains of virus in a chicken egg, let them replicate for a while, then comb through the resulting mishmash of different viruses for one you need for the vaccine.

"WHY DIDN'T WE see this coming? Why aren't we prepared?"

"Because, Mr. President, nobody thought that an H1 virus that had been circulating for years among humans could turn into a pandemic. We were expecting the next outbreak to be an H5 virus—bird flu. By the time we knew about this H1N1 outbreak, it was already a pandemic. And because it's a new strain, a vaccine could take months.

"We need to prepare, Mr. President, for mass panic. There will soon be a shortage of hospital beds, respirators, and medical expertise. And then people will start dying.

"There will be a need for mass burials."

EVERY HEALTH DEPARTMENT of every major city, as well as the Department of Health and Human Services and the Pentagon, has a document that it hopes never to have to dust off. It is called the Mass Fatality Plan. It outlines what would need to be done in the event of a disaster such as the next killer influenza epidemic. It assesses the "surge capacity" of local hospitals—how many beds are typically available, how many could be made available by sending some patients home and postponing elective surgery. The New York City area, for instance, has a limited supply of respirators—devices to help people with fluid in their lungs to breathe, which would be essential in a flu emergency. In a 1918-style flu pandemic, these respirators would be at a premium, and hospitals would be forced to ration their use. That means many people would have to be left to drown as their lungs filled up with fluids.

They would have to stay home, since hospital beds would also quickly be filled up. A surge in influenza intakes would quickly overwhelm the nation's hospitals. Currently there are just under a million hospital beds in the nation. Most hospitals operate at about two-thirds capacity—which comes to roughly three hundred thousand extra beds. An influenza outbreak that sickened 1 percent of the nation at once would use up all spare capacity. (That's assuming, of course, that these beds are distributed uniformly, which of course they aren't.) If 10 percent of the population marched into emergency rooms, there would be pandemonium.

Sending patients to nearby towns and cities would be pointless because they would likely have their own shortages. A pandemic flu would seem to strike everywhere at once.

By mid-May, hospitals in New York City and several other major cities are at the breaking point. Then, as if by a miracle, the number of casualties begins to level off and decline. For reasons that scientists don't fully understand, influenza viruses wane in the spring and summer months. The flu season is nearing an end.

But the flu that rages briefly through the spring will come back in the fall. What form the virus takes is impossible to predict: it may morph into something less deadly, or it may turn even more deadly. Only time will tell.

Over the summer, vaccine makers labor to prepare their concoctions, rushing to bring out a vaccine as quickly as possible. By July, officials are certain that they'll be able to make a vaccine, but they are still months off. Because the flu is new, they need to replicate the virus in the lab, make an attenuated form that won't harm anyone but will stimulate the immune system to respond, and then test it. What they don't know is how many shots it will require: two shots per person take longer to produce than one. By August, it's apparent that the vaccine won't be ready before late October, at the earliest. Since most people will need two shots, there won't be enough to vaccinate the entire population until

January. To prepare for a rough few months, the federal government moves to stockpile antivirals such as Tamiflu.

Meanwhile, health officials are anxiously watching the progress of the virus around the world. It's grim. In South America, millions of people are getting sick, and hundreds of thousands are dying. Australia declares a national emergency and appeals for hospital equipment from nations in the northern hemisphere, but helping out is politically difficult if the equipment and medicines will be needed at home. As September nears, the White House decides to order schools to remain closed.

Stock markets around the world, already depressed, greet the news of school closings with a mass sell-off. Prices plummet. Trade comes to a virtual standstill. Even before the virus hits for real, an economic crisis sets in. The only stocks that rise are those of mortuary services.

A trickle of people are allowed to cross national borders, but travel is significantly limited. Airlines are filing for bankruptcy.

By the beginning of October, hospitals are no longer accepting patients. News reports tell of the sick dying in reception areas.

As the dead pile up, some states close their borders to keep medical refugees out. Shipping comes to a stop, causing shortages of heating fuel, food, and medical supplies. Because people are afraid to congregate, absenteeism begins to affect businesses of all sorts. Electrical utilities experience periodic outages due to lack of maintenance staff, who are either sick or home scared.

The army and the National Guard begin to take over the distribution of basic food and medical supplies, and the burying of the dead. As the death toll exceeds 10 million by November, corpses start piling up. A backlog of the dead develops. The federal government mobilizes the army to dig mass graves. It's a matter of public health: each dead body is a reservoir of disease, waiting to find another host. The army issues directives on attaching identification to the bodies of loved ones so they can, at some later date, be reinterred in family plots. Refrigeration trucks are commandeered to keep bodies from decomposing.

IT'S HARD EVEN to imagine the effect mortality on the order of a severe pandemic would have on our modern world. You would have to go back to the Black Death that swept through Asia and Europe in the fourteenth century to come up with an analog.

The Black Death struck a very different planet than the one we now inhabit. News traveled slowly back then, and so did people—the trip from

Crimea to China, where the disease first emerged, took about twelve months over the mountains and along the hardscrabble roads. The strain of bacteria that caused the plague took several years to make its way across the Eurasian continent, hitchhiking on rats and their fleas.

When it arrived, though, it seemed like Armageddon to those who lived through it. In Europe, it killed about a third of the population—as much as 60 percent in some places. The population of China dropped 50 percent.

Observers in Europe reported corpses in the street

> packed like "lasagna" in municipal plague pits, collection carts winding through early-morning streets to pick up the previous day's dead, husbands abandoning dying wives and parents abandoning dying children—for fear of contagion—and knots of people crouched over latrines and sewers inhaling the noxious fumes in hopes of inoculating themselves against the plague. It was dusted roads packed with panicked refugees, ghost ships crewed by corpses, and a feral child running wild in a deserted mountain village. For a moment in the middle of the fourteenth century, millions of people across Eurasia began to contemplate the end of civilization, and with it perhaps the end of the human race.[10]

The 1918 flu pandemic was not much better. The disease first showed up in army installations during the world war, dropped off over the summer, as flu tends to do, and roared back in the fall. Philadelphia, one of the first cities hit, saw 1,650 cases in the first day. By day five, 2,600 people had died. Within a month, the death toll had risen to 11,000. Doctors and undertakers were inundated. The virus swept through every city in the world, every rural district.[11]

The world of 1918 was slower than the world is now. It took weeks to travel from Europe to America, and not many people made the trip. Influenza viruses now have new possibilities for mischief, and their evolution has only begun to explore them. A creature's habitat, to a great extent, dictates its potential. The writer John Kelly estimates that pestilence on the scale of the Black Death would claim almost 2 billion lives.

As bad as it sounds, it's not the worst thing that could happen.

Many people argue that there is something qualitatively different about the trajectory of human civilization in the twenty-first century from the early twentieth or the fourteenth centuries. This is the argument that we're getting close to some kind of fundamental limit—of earthly resources, of habitat, and of the complexity of a planetary ecosystem that we seem unwittingly to have taken charge of. Many scientists believe we are currently on the verge of a "mass extinction event."

A true mass extinction would make the Black Death look like a mild recession by comparison. It is something that *Homo sapiens* has never experienced, and will only ever experience once.

CHAPTER 2

Extinction

THE LARGE BLUE BUTTERFLY IS A gorgeous creature. Generations of Britons grew up seeing these blue-and-white creatures in the countryside by the thousands in the summer months. They flitted among the meadows to the delight of walkers and nature lovers and farmers and parents, who used their appearance to teach their children that gossamer beauty could emerge from something as hideous as the lowly caterpillar. In the 1970s, though, Large Blues became increasingly hard to find. Biologists declared that the creatures were dying off, for reasons that remained a mystery. Many Britons, goaded by the tabloids, began to blame the poor butterfly catchers. For a time, public opinion being what it was, nature lovers were becoming increasingly wary of being seen out in the field with their nets and their collection boxes, and it was a wise precaution, after an outing, to stash the equipment in the boot of the car before walking into the pub. In 1979, scientists declared the Large Blue to be extinct in Britain.[1]

Funny thing about this extinction, though: nothing seemed to have caused it. The butterfly's habitat—the meadows of the British countryside—hadn't appreciably changed in decades. By all appearances, the butterfly should still have been thriving, yet it was gone from the windy isle, though it lived elsewhere in Europe.

To crack this puzzle, scientists studied the creature in all its stages of development, from embryo to caterpillar to cocoon, right up to the point when the butterfly emerged and spread its wings. They found something odd: the caterpillar, before making a cocoon, would leave the thyme and marjoram plants it feeds on and begin to crawl around in the dirt, masquerading as another species entirely: the larva of a particular species of red ant, *Myrmica sabuleti*. Although the caterpillar actually looks something like an ant larva, the disguise is largely biochemical. The caterpillar has a gland that secretes a sweet fluid that attracts the ants, which feed off the fluid. Then they take the caterpillar back to their nest, embracing it as one of their own. The caterpillar hibernates for a few weeks in the ant nest, then, when it's good and ready, turns on its hosts and feasts on their eggs and larvae. A single caterpillar will eat more than a thousand ant larvae and grow to one hundred times its original size, right there in the ant colony. The caterpillar then spins a cocoon, pupates, and emerges, six weeks later, from the ant tunnels as an adult Large Blue. Not only do the ants put up with this, they often escort the butterfly to the surface and help ward off predators until its wings are dry and ready for flight.[2]

But the caterpillar's ruse only works completely with this one species of red ant—and there's the rub. That one species had also disappeared at about the same time the Large Blues did. It went unnoticed, however, in part because the caterpillars continued to shack up with a close relative of *M. sabuleti*—another species of ant, *M. scabrinodis*, that is identical in almost every respect, though not in one crucial one: it was not as easily fooled by the caterpillar's inpersonation of an ant larva. *M. scabrinodis* would adopt the disguised caterpillar, but they got hip to the ruse quickly—too quickly for the caterpillar's good. Before it had a chance to grow, spin, and pupate, the ants realized that they had taken an impostor into their midst, and they would turn on the caterpillar, feasting on it before it could feast on them. So when *M. sabuleti* disappeared from Britain, the Large Blue butterfly soon followed.[3]

But now there was a new problem. Where had the ants gone, and why? It didn't take long to find the culprit—human activity. It's tempting to think it had something to do with the construction of tracts of McMansions or theme parks or shopping malls, but it was more complicated than that.

Biologists found that *M. sabuleti* had been taking advantage of a niche in British-countryside ecology that had begun to open up five thousand years ago, with the introduction of sheep to the island. Before sheep were commonplace, the Large Blue must not have been quite so abundant in the fields. British farms have been around for so long that the Britons have come to think of the hedgerows and crop plots as an inevitable and natural part of the landscape. But ancient humans cut down forests to open up the land; and in those places where there weren't crops, grasses grew, and sheep grazed, keeping the grass short with their constant nibbling. *M. sabuleti*, in turn, is nature's adaptation to the new ecological niche of short grass. The Large Blue also adapted perfectly to the profusion of red ants. Subsequent generations of townsfolk grew up assuming, perhaps unconsciously, that the Large Blue had always been there and always would be, that it was a staple of nature, as steady as the sun and the rain.

But times change. (They'd been changing even before the industrial revolution and the Internet revolution and the biotech revolution turned change into a rapid-fire constant.) Farms, even in the smallish country of Britain, grew bigger, and more intensive. Many former sheep-grazing lands became suburbs or exurbs, where the grass is left largely alone. It grew, the red ant's niche closed up, and the Large Blue lost its place as well.[4]

Such is the fragility of species interdependence.

This particular story has a happy ending, however. After the biologists solved the puzzle, conservationists went to work to reintroduce the butterfly from Sweden and bring cows to the pastures to keep the grass short. Large Blues began to thrive. Now when walking along the ancient pathways and over the fences of the British countryside, one once again sees a profusion of Large Blues.

The plight of the Large Blue is a story in miniature of the web of life, of the interdependence of species, of how things are connected in ways that are hard to anticipate and imagine. It is a good place to start this discussion of extinctions. The island of Britain is relatively small, and the Large Blue and *M. sabuleti* and the British sheep are just three species in an exceedingly complex dance of interdependency among many thousands of species of creatures large and small. Imagine for a moment these links extending to other animals and plants, and the bacteria and insects and fungi. Now imagine weather and climate and oceans mixed in, throw in a few asteroid impacts and gigantic lava eruptions, and you can begin to get a glimpse of the ebb and flow of the planet's geology, ecology, and the species that depend on these things and on one another.

The plight of the Large Blue is a parable of how nature isn't always the steady thing that we think it is. It is constantly shifting, fluctuating, changing. Sometimes the changes appear random but stay within certain boundaries—a kind of steady state, with some small variations. Sometimes, though, big changes take place suddenly and ripple through the entire chain of creation. Sometimes small changes become big changes, and big changes add up to gigantic disruptions—catastrophic ones.

But we're getting ahead of ourselves.

NOBODY KNOWS HOW many species there are in the world right now. Scientists have named between 1.3 million and 1.8 million species; estimates of the total number run from 3.6 million to more than 100 million.[5] Of these, more than a million are insects; about twenty thousand of these are butterflies.[6] A few million years ago there may have been three or four hominid species at any one time. Two of those species eventually died out, leaving Neanderthals and *Homo sapiens*. Neanderthals disappeared about fifty thousand years ago, and now *Homo sapiens* is the last of its kind. (The Neanderthal who lives in the house next door only seems like a separate species, but he isn't really.)

Before Darwin's day, the notion that species were some immutable constant, like the stars above and the ground underfoot, was conventional wisdom. Darwin, though, dispelled that idea. Instead, his idea was that species formed to adapt to changing circumstances. Parents got by somehow, but their offspring differed slightly, sometimes in ways that worked to their advantage and sometimes to their disadvantage. Darwin observed the adaptability of finches in the Galápagos—small birds that existed in abundance on these islands in the 1800s. Darwin observed dozens of different species, with beak sizes adapted for island niches.

Darwin's finch observations were later given further scientific study. A baby finch may be born with a slightly longer beak that makes it better able to eat a certain kind of berry that is suddenly more prevalent, perhaps because the rain fell more frequently and suddenly created the conditions perfect for this particular food source. So the baby finch thrives.[7] You begin to see a very different picture of the canvas of life after Darwin—not a still life, but rather an ever-shifting kaleidoscope of changing forms. Species come and species go. On Earth, life is the most changeable of things, and it adapts to the forces around it.

Sometimes, though, change happens all at once, too quickly for species to adapt. Many biologists believe that is happening now. They believe that humans are changing the planet at a rate too quick for many of the planets' species to keep up. Eventually, they fear, species will die off suddenly en masse, triggering a cascade of extinctions, which could have a devastating effect on the planet's intricate web of life. It's the Large Blue phenomenon multiplied throughout all the ecosystems on the planet. It's nature's way of wiping the slate clean to begin again. It's happened at least five times since life began on the planet about 4 billion years ago.

Is it happening now? The question is important because if the answer is yes, the likelihood that humans will survive is slim.

A closer look at a few of these extinction events of the past will give some clues.

GOD INSTRUCTED NOAH to build an ark and fill it with two of each kind. By the time the waters receded, all the earth was barren, and it was time to start anew.

Thanks to Noah's assiduousness, he preserved the diversity of species that had existed before the flood—he took two of each kind, and the earth was replenished. (Strictly speaking, Noah did some big damage to genetic diversity, of course.) A true mass extinction is not so kind. In the 4 billion years or so that life has existed on this planet, nature has periodically wiped the slate clean, so to speak, and killed off a great portion of the flora and fauna in existence, never to appear again.

Most of Earth's extinction events involved creatures that no one but a paleobiologist could love. We're talking mainly about microbes and little fishy things that resemble the kinds of shelled creatures you cut your feet on when you're wading the intertidal pools at the beach. The most famous exception, of course, is the most recent mass extinction event—the one that killed off the dinosaurs 65 million years ago. Because of these big, charismatic reptiles, many schoolchildren these days become acquainted with the story of the big meteorite that landed one day in what is now the Yucatán Peninsula of Mexico and ended the days of *T. rex*.

A few billion years earlier, a meteorite like the one that killed the dinosaurs would have been commonplace. The solar system was full of asteroids, and impacts happened with great frequency. Then things calmed down a bit, and life was given some space to form and evolve into increasingly complex forms. But a mere 65 million years ago there weren't that many big objects left—most had already slammed into one another or the major planets. The meteorite that killed the dinosaurs was a rarity, and for them it was a disaster.

The end of the dinosaurs, known as the K-T extinction event, shows how dramatic a mass extinction event can be. It's the one we know most about, and since it's the most recent, we can paint it with the most vivid detail. The great die-off that occurred 250 million years ago, in the Permian era, is in some ways more interesting, but it involved creatures that would not make good characters in a movie. For sheer drama, the dino die-off takes the prize.

The meteorite that took out the dinosaurs was big—it would have been about nine miles in diameter, probably an asteroid or a comet—and struck Earth head on with a force of 100 million megatons, the equivalent of about 1 billion Little Boy bombs (the one dropped on Hiroshima).[8] The meteorite formed a crater 110 miles wide. Geophysicist Glen Penfield found the crater in the Yucatan in Mexico in the 1970s while prospecting for oil and spent years trying to confirm the discovery. At about the same time, father and son geologists Luis and Walter Alvarez, on a field trip to Italy, chipped off a sample of stone from a layer in the rock that dated to 65 million years ago. The stone contained the chemical element iridium, a silver-white, brittle substance. Subsequent fieldwork found the same iridium deposited in geological strata of the same age in all four corners of the globe. Many prehistoric events are shrouded in ambiguities of incomplete evidence, but not this one. It was as clear-cut a case of classic cause and effect as you could ever hope to find in studies of past natural events.

The discovery of the Chicxulub crater completely changed scientists' concept of the dinosaur extinction. Suddenly all sorts of other hypotheses—such as the theory that the dinos were done in by the eruption of a supervolcano—became irrelevant, and scientists were able to piece together a narrative of that fateful moment when the dinosaurs, rulers of Earth, fell to oblivion in the span of a long weekend.

The meteorite impact was huge. It would have been visible in the sky as a bright star days and perhaps weeks before impact. Had it happened tomorrow, NASA scientists might alert us to the event a few weeks ahead of time, calculate the trajectory of the object, and demand an audience with the president and go on talk shows and news programs in a campaign to get funding for an anti-asteroid program. Perhaps, if we were really lucky, the NASA folks would have gotten a bead on the meteor a few months ahead of time, giving us at least a prayer of launching a nuclear-tipped missile that might have been able to break up the meteor into smaller parts that would fall relatively harmlessly in the atmosphere, or knock it off course enough to miss Earth. U.S. Congress mandated in 2005 that

NASA find nine of every ten asteroids that could strike Earth and do some damage,[9] so that we could perhaps have years of warning before a potential impact and prepare a plan for dealing with one. But with a measly $4 million budget, NASA's asteroid watch is likely to fall short.[10] Still, that's more than *T. rex* had.

The dinosaurs were probably oblivious to the slowly brightening star in the night sky, but in the hours before the Big Event, they might have registered some slight bewilderment at the growing light, which seemed to come out of nowhere. At the fateful moment, animals within a radius of thousands of miles would have heard the thunderclap of the meteorite breaking the sound barrier, screaming across the skies like the face of death itself. Creatures on other continents would have felt the rumble as the impact rolled and reverberated throughout Earth's crust, ringing the planet like a bell.

The impact would have been like nothing *Homo sapiens* has ever felt. The meteorite that in 1908 fell over Tunguska, in Siberia, exploded with an impact of ten or fifteen megatons, flattening trees and causing forest fires for hundreds of miles. But that was ten million times smaller than the Chicxulub meteorite, which is why it exploded in the air, never reaching the ground. Between the forest fires and dust from the impact, the Chicxulub meteorite kicked up a cloud of dust that spread throughout the globe, causing a dimming of the sun.

By odd coincidence, the Chicxulub hypothesis emerged during the Cold War, when much of the world had already been girding for a nuclear disaster. In 1982, writer Jonathan Schell published *Fate of the Earth*, in which he described, in grisly detail, what a nuclear-weapon Chicxulub-like explosion would do to the planet and to the creatures that live on it. Much of what Schell described—the nuclear winter that would cause temperature drops and kill off crops and so forth—was similar to the dinosaurs' nightmare, with one big exception. The death cloud in Schell's narrative held significant amounts of radiation. None of the natural impacts or big volcanoes that played roles in past mass extinctions delivered that kind of nuclear radiation.

What they would have created, though, is acid rain—a lot of it. Sulfur exists in abundance in the lava that flows from Earth's mantle. We know this, in part, because we've seen a smaller version of it in our lifetime: the eruption of Mt. Pinatubo in 1991. Pinatubo spewed 20 million tons of sulfur dioxide into the atmosphere,[11] which rose into the skies and interacted with the upper atmosphere to form a blanket of aerosols—a mix of liquid droplets, smoke, and dust. The aerosols attract moisture and eventually promote the formation of clouds, which in turn produce rain that is highly acidic. This planetary haze reflected the sun's energy back into space, causing a drop in average global temperatures of half a degree centigrade that lasted about two years. The aerosol covering also severely diminished Earth's ozone layer.

The Chicxulub impact probably blocked out a great proportion of sunlight for years, with effects lingering over a decade. It obliterated the planet's ozone layer, allowing ultraviolet rays to penetrate the atmosphere and reach the earth's surface. The sunlight was severe enough to cause a sunburn after five minutes of exposure and severe burns in a few hours. The acid rain delivered yet another blow to plant life. Only the hardiest of creatures survived these conditions. Half of all species, scientists estimate, went extinct. The web of life took a big hit.

If a Chicxulub-type impact occurred during our times, the disruption would be difficult to fathom. What the impact didn't destroy, the forest fires and the aerosol pollution and the ozone depletion would. Crops would not survive such conditions. Famine would ensue, and people who weren't incinerated by the impact would die due to starvation and burns. The world would instantly turn into a living hell.

But these aren't the kinds of disasters we're concerned with here. The destruction of our planet by meteorite impact is an externality: it simply happens or doesn't, and no crack team of rock drillers and nuclear engineers is likely to save us. (Though a decent NASA asteroid program might be a good insurance policy.) The more interesting calamities are those of our own making.

We humans wouldn't be here if not for the Chicxulub impact. For all its destructiveness, by eliminating the dinosaurs at the top of the food web, it created space for a new dominant order: mammals. Mammals had been scurrying around the feet of the dinosaurs for years, a mere nuisance to their bigger cohabitants. Small mammals, however, for reasons that are still poorly understood, better withstood the aftermath of the Chicxulub impact and, with the big reptiles gone, found themselves suddenly without nearly as much competition for food. But this would have taken many millions of years to become obvious to an onlooker. Indeed, from the standpoint of the individual, this kind of evolutionary revenge is almost always served cold: most creatures of any species to benefit from it surely perished in the cataclysm.

The extinction of the dinos is certainly interesting and presents a possible scenario for the extinction of the human race, but it is not nearly as interesting, for our purposes, as three prior extinctions—extinctions that took place 252 million, 600 million, and 2,400 million years ago. For all the dullness of life on the planet in those days, the causes of those die-offs more closely approach the one we may currently be on the verge of. (Or perhaps we're in the middle. Biologists who observe the living world, and who see species disappearing at an alarming rate, tend to think that a mass die-off is under way.) It turns out that microbes can effect the planet's geochemistry in ways similar to how we're changing our own right now.

So let's continue our trip backward in our time machine, to the Permian era, and the greatest case of mass death the world has ever known.

LIFE ON EARTH 252 million years ago was considerably different from what it is now. Animal life in the oceans consisted mainly of creatures anchored to the sea floor—sponges and corals and brachiopods. Typical of the life-forms at the time were the echinoderms. These creatures lived atop a stalk of circular calcite plates attached to the seafloor and extended their arms out in a circle to filter organic material out of the water. Bryozoans, another type of filter feeder, lived in colonies that formed a kind of fan that stretched up from the seafloor. There were sea lilies and ammonoids, distant relatives of the nautilus. These stationary animals greatly outnumbered the trilobytes, fish, snails, and sea urchins that were ambulatory. There were precious few squidlike creatures, but plenty of foraminifera, single-celled protozoans that can swim. The world's continents were combined into one large land mass, Pangaea, on which lived a variety of reptiles, amphibians, and fixed-wing insects much like modern-day dragonflies.[12]

This is what the world was like before the event that would obliterate 90 percent of all species and would put an abrupt end to the Permian era.

In the modern world, this underwater ecosystem has long since been buried under the sediments and scattered among the world's continents. One piece in particular—a portion of a shallow sea that lay along the equator, a hot spot of biodiversity that may have been the Permian era's version of the Great Barrier Reef of Australia—was preserved and buried under layers of rock. As Pangaea broke up and drifted into the continents we know today, this hot-spot-in-rock drifted, too. It wound up in Texas, where geological forces pushed it upward. Thousands of years of rainfall began to wear it away, and most recently, the fast-flowing McKittrick Creek carved a slice like a hot spoon in ice cream, exposing the record of the ancient Permian reef for all to see.

Like many paleobiologists before and since, Doug Erwin made the trek in the 1990s to Texas and stood at the edge of the creek, at the bottom of what was once an ancient seabed, and looked upward at the steep escarpment. He filled his canteen with the cool water and started up the Permian Reef Trail, slowly climbing switchbacks, twisting between juniper trees until he was twelve hundred feet up. Then he came upon huge blocks of limestone, formed from the accretion of discarded shells of long-dead creatures. When he'd see what he thought might be a fossil, he'd open his canteen and pour water over the rocks, which caused the outlines to appear as if in bas-relief. "Some rocks consist of nothing but half-inch-long grains of rice: the skeletons of single-celled organisms known as foraminifera." Clamlike creatures called brachiopods are also abundant, but they wouldn't have been very satisfying served with garlic on linguine: instead of meat these creatures had filaments that filtered water for organic matter—bits of dead creatures and single-celled organisms.

Then Erwin began to look at another spot, hundreds of miles away, near the Green River in Utah, where fossils in a stratum of ocean sediment only a few million years younger showed a world of an almost entirely different order—the relatively barren times of postextinction life.

Fossils near the Green River in Utah are much less abundant than those in Texas. Whereas Texas has fossils of a hundred species of snails, Utah holds only nine or ten. Meanwhile, the Utah site shows a relative abundance of species that got lost in the pre-extinction profusion of life forms. Erwin found fossils of a type of scalloplike clam, called *Claraia*, in the tens of thousands. Starfish, once rare, were suddenly common.

The work of paleobiologists in Texas and Utah, and confirmed in China, South Africa, Iran, and Italy, where pieces of the ancient world were also preserved, showed that more than 90 percent of all the species on the planet died off about 251 million years ago. In a span of time that seemed almost impossibly short in geological time, life on Earth had been decimated, and a new order of life had begun to take shape.

The fossils in Texas and Utah represent so different a world from ours and from one another that John Phillips, a mid-nineteenth-century geologist, took them as evidence for two separate creations. Phillips had no way to judge how long ago these different "creations" occurred or how much time had passed between them. But he was the first geologist to recognize the grand patterns in the fossil record of life on Earth, and he divided the history of animal life into broad time periods. We now know these as the Paleozoic, up to 250 million years ago; the Mesozoic, up to 65 million years ago; and the most recent period, the Cenozoic. Between these periods you often found steep drops in the number of species in the fossil record, followed by new explosions of species.

Phillips never accepted Darwin's theory of evolution. He didn't believe that these periods were linked by a common thread, but thought instead that each time the planet was wiped clean and life began anew from constituent molecules in the primordial soup. Subsequent study of the fossils showed that he was wrong about that—some species persisted through extinction events and sometimes began to thrive anew, the way the mammals thrived after the dinosaurs were killed off, and, as we'll see, other creatures thrived after the Permian extinction. Phillips focused on figuring out exactly when each of these periods occurred, and how to place fossils in the geological record. His work created a basic timeline for extinctions on this planet, which is still in use today.

As Erwin studied the Permian extinction, he was surprised at how quickly it seemed to have occurred. The fossil record is, at best, like a thick Magic Marker—it doesn't draw a fine line. You can see plainly in one geological stratum tons of fossils, and in another layer hardly any. But what happens instantly in geological time doesn't tell you much about what might have happened over the human time scale—over the course of a generation, or a lifetime, or just over the horizon, within sight of your children's grandchildren. Erwin might have mused to himself at a private moment about scenes of devastation and destruction, but the thought didn't enter the professional side of his brain that anything could have happened that you could make a movie out of.

But he and his scientific colleagues could tell, despite the fuzzy fossil record, that the Permian extinction was quick—perhaps 5 to 10 million years, which is an instant in geological time.[13] In the late 1990s, archaeologists fanned out to southern China, Italy, Austria, and South Africa to get a bead on the causes and the quickness of the event. Access to fossils was difficult to come by at first. There were plenty in southern China, but many were out of reach of archaeologists. Gradually, though, more fossils from the period came to light. Finds in Greece and China suggested the die-off happened as quickly as within a million years or so. Some species, though, seem to have tapered off over millions of years—such as the snails that Erwin studied, whose decline seems to have preceded the die-off. In short, the scientists were getting some mixed signals.

In the past few years, however, they have narrowed the Permian event to less than two hundred thousand years—a veritable New York minute—based on close cross-correlation with the myriad fossils found throughout the world. Erwin says it is possible that the extinction event happened in a few tens of thousands of years, but there is no way to know from the fossil evidence. That may not seem quick, but it brings the duration of the event into a time period with human significance. Agriculture began ten thousand years ago; one hundred thousand years ago *Homo sapiens* began its migration out of Africa.

Erwin, as a paleontologist, won't speculate on what might or might not have happened in the Permian extinction; he survives on evidence alone. His latest shows that the die-off happened in something like ten thousand years, but that doesn't mean it actually took that long. There's no way to know right now. There is some reason to think that the Permian extinction could have happened quickly, perhaps in the span of a human lifetime, or perhaps in a bad dino-type nightmare; this is speculative, and we'll save it for the next chapter.

Yet we can begin to paint a picture of what happened to the planet 251 million years ago. It started most likely with a volcanic eruption. This most certainly happened on what is now the Siberian steppes, where layers of solidified lava are trapped in the ground. The geological record shows some four dozen or so separate lava flows, some of them as thick as thirty-seven hundred meters—more than two miles—taking up at least 1.5 million cubic meters. That is one gigantic outpouring of lava. The Mt. Pinatubo eruption, by comparison, as big as it was, didn't produce a drop of lava—most of its global impact came from the fountain of ash it spewed into the atmosphere. It's impossible to tell from the geological record how much ash the Siberian volcanoes of the Permian era threw into the atmosphere, but it must have been huge. Those eruptions are considered to be the biggest volcanic event to occur on land for the past 500 million years. They lasted for a million years, and their lava flows produced what are now the vast Siberian traps, equal in land area to Western Europe. It's certainly plausible that their global effect on climate and weather caused the extinction of many species.

Many species—but what about almost all species? That is a tall order. It's difficult to wipe nine of every ten species off the face of the planet, which is what happened in the Permian extinction. Volcanism, even on the order of these tremendous Siberian eruptions, would not have done the trick.

The real culprit, it turns out, may be lying in plain sight. Today Siberia is home to the world's largest deposit of coal. The Chelyabinsk basin is big—175 billion tons—but it was a lot bigger before the volcanoes started erupting in the end Permian days. The lava flowed and flowed and eventually reached the giant coal reserves. The hot rock would have vaporized much of the coal, turning it from a solid into methane gas and carbon dioxide.

You will recognize these gases from the newspapers. Methane is what comes from cow farts, rotting wood, and melting permafrost. It is a potent greenhouse gas—far more potent than carbon dioxide, which is the prime suspect in climate change in part because of the sheer volume our tailpipes and smokestacks produce. But a huge and sudden release of both of these gases from the coal basin in Siberia—*that* is big enough to plausibly trigger an extinction event. It caused a vast and sudden change in the planet's geochemistry—similar to the drastic changes humans are making now. Temperatures would have climbed steeply, especially with all that methane in the atmosphere. Methane produced in that volume would have triggered a temperature rise that killed off a bunch of species outright. Then, after a few years, it would have degraded into water and carbon dioxide, which lingers in the atmosphere for many decades.

To make things worse, the volcanic eruptions most certainly created an intense acid rain for many years afterward. Volcanic eruptions, as NASA's satellite images of Mt. Pinatubo have revealed, release sulfur into the atmosphere in massive quantities. The raindrops that form on it are highly corrosive and toxic.[14]

What's certain is that 90 percent of all species on Earth disappeared. Land and sea became bereft of life save for a few hardy survivors. The sea urchin did quite well, going from a minor player before the extinction to a major presence in the new oceans. Some other species here and there hung on—ammonoids and gastropods, the conodonts and the primitive chordates, whose easily preserved mouthparts serve as important markers of time to paleobiologists today. A few species of clams did well—particuarly *Claraia*, which began to thrive and soon spread throughout the globe. Over a few million years, the oceans began to come back. The age of the stationary, floor-dwelling marine creatures had come to an end, and the oceans began to bear some resemblance to what we know today. The real winners of the extinction event were those creatures that could swim—the squid and the trilobytes. More species developed the ability to burrow, suggesting that more predators were swimming around looking for tasty morsels. The modern seafood-restaurant menu began to take shape.

Life immediately after the Permian extinction was desolate, but the next few million years saw an explosion of life-forms that had different skills—for instance, insects that could retract their wings. It replaced the fixed life-forms of the seas with new ones that could move and give chase and burrow beneath the ground. Had humans been among the 90 percent of species that didn't make it, we'd hardly see this as a positive step. But it was nevertheless an example of nature's own nuclear option: wiping the slate clean and starting almost anew.

THE PERMIAN AND K-T extinctions are not the only games in town. As we said, five or six earlier extinction events occurred during the course of life on the planet. Each time, these events seemed to clear the way for something new and different and ultimately beneficial, when you look at things from the standpoint of geological time—which is to say, over millions of years.

The Cambrian explosion about 600 million years ago, at the dawn of animal life, was a rapid proliferation of new life-forms that was unprecedented at the time and has never been equaled since. The abruptness of this radiation of life-forms bothered Darwin greatly, because it didn't seem to fit into his notion of gradual evolution from natural selection. In recent years, scientists have sought to piece together what happened, and it doesn't contradict Darwin so much as add complexity.

Before this explosion in new life, there is some evidence of a smaller proliferation of life and subsequent die-off called the Ediacaran. Basically, the Ediacaran episode resulted in the rise of worms and bloblike, soft creatures which may then have dropped off suddenly, for reasons that are obscure, perhaps clearing the way for the Cambrian explosion, which created, in a mere 10 million years, the new body blueprints we recognize today—heads and arms and legs and digestive systems and nervous systems and cardiovascular systems. A few of the Ediacaran worms survived.

So life after a mass extinction isn't so bad, if you're willing to wait a few million years. Or you're a worm.

• • •

THE PERMIAN EXTINCTION is a kind of cautionary tale, a preview of how climate change can annihilate life on Earth. Human activity, specifically the release of greenhouse gases into the atmosphere, appears to put us in a similar predicament. We are at risk of becoming the second species to affect the environment so profoundly that we cause an extinction event, in effect wiping ourselves off the planet.

If we are the second, then what was the first? The Great Oxidation Event, 2.4 billion years ago, is also a case of an organism changing the world's chemistry in profound ways, threatening the lives of other creatures around it. The villain is a little creature called the cyanobacterium. Also known, colloquially, as pond scum.

Two and a half billion years ago, life had evolved from bits of genetic material (something like viruses) to genetic material contained in cells (bacteria). The only bacteria that could manage on Earth at the time were anaerobic, meaning they did not require oxygen to live. In fact, oxygen was toxic to these bugs.

Anaerobic bacteria still exist today, but back then they had the upper hand. Earth's atmosphere had little oxygen in it. But then something happened. Some of these anaerobic bacteria evolved to new forms—a precursor to modern pond scum. And in the normal course of their daily lives, they started producing oxygen.

At first none of the anaerobic bacteria noticed the oxygen their pond-scum neighbors produced. For years, the two life-forms sat right next to one another. For most of that time, nothing much happened. Whatever oxygen went into Earth's atmosphere was quickly absorbed by iron in the surrounding rock—the iron, in other words, rusted. Earth in those days had a lot of pure iron that hadn't yet oxidized. For hundreds of millions of years, the aerobic bacteria kept producing gas like the flatulent rider on a commuter train, and Earth kept rusting and rusting, until it could rust no more. All the iron had been used up.

The oxygen kept coming—the pond scum were doing quite well—but it had nowhere to go. It just stayed in the atmosphere. Eventually the pond scum gassed the anaerobic bacteria to kingdom come—at least the ones that couldn't scuttle under some crack or hide in a deep-water vent, away from the oxygen-rich air. "From the standpoint of a microbe 2.4 billion years ago," says Erwin, "the onset of the introduction of all this oxygen was a horrible event, because it changed the environment completely. Anaerobic bacteria had to go hide in the mud or something. Oxygen would have been a toxin to a lot of microbes that were adapted to a very different environment."

That the pond scum gassed the anaerobic bacteria into submission is interesting enough. What makes this story especially compelling is that it may have happened suddenly and catastrophically. Evidence of a decisive pond-scum victory has come to light only recently. The reasoning is somewhat arcane, involving the existence of oxygen and sulfates in the oceans and rocks. Scientists used to think that the presence of hydrogen sulfide in the oceans prior to the extinction meant that oxygen had been accumulating slowly in the atmosphere over millions of years. Oxygen in the air interacts with rocks, causing them to rust, but also producing sulfates. The rain and the rivers wash the sulfates into the sea, where bacteria turn them into hydrogen sulfate. Ergo, the presence of hydrogen sulfate in the oceans seemed to constitute evidence that the air contained high levels of oxygen over millions of years, and that pond scum came to dominate gradually. But that scenario may not be correct.

Timothy Lyons, a biogeochemist at the University of California, Riverside, came up with an alternative explanation for how all this hydrogen sulfide could have accumulated in the oceans. Volcanoes might have spewed the sulfates instead, which bacteria then converted to hydrogen sulfate. The atmosphere, in this scenario, stayed oxygen-poor up until the last moment before the extinction. Lyons found supporting evidence for this hypothesis. He found pyrite from that early period, which generally forms in oxygen-poor environments. And he found little evidence of molybdenum, an element that is a prerequisite for algae. "The scarcity of molybdenum in rocks deposited 100 million years earlier . . . suggest that cyanobacteria were probably struggling to produce oxygen when these rocks formed," said Clint Scott, a graduate student of Lyons' who coauthored a paper on the findings.[15]

The die-off happened quickly, at least by the standards of paleobiology. Scientists know about the plight of these microbes because they find fossilized remains of them. Like Erwin's walk through the Permian Reef Trail, they find one kind of fossilized microbes in one layer, then all of a sudden they find another kind of microbe in another layer. Things are all anaerobic, then suddenly they're aerobic. It seemed to happen quickly, but how quickly? Over a million years or a decade? It's difficult to judge.

For us, the triumph of pond scum was a good thing. Oxygenating photosynthesis became the basis for the plant kingdom we have today. We rely on this system to absorb carbon dioxide and refresh the atmosphere with oxygen.

Today we're the pond scum, pumping out not oxygen but carbon dioxide, filling up the oceans and the atmosphere, geochemically transforming the planet.

Are we now engaged in bringing a new mass extinction event upon ourselves? If so, you might call it the Holocene extinction, because we are currently in the Holocene era, which began ten thousand years ago. That was about the time hunter-gatherers began to settle down and tend their farms, which freed up some spare time to make cities and write books and invent the Internet. From the standpoint of the paleobiologist 250 million years from now, will the Holocene be known as the beginning of yet another grand ending? Or a new era we might call the Anthropocene?

Whether that is, strictly speaking, true is a matter of some debate. Biologists, troubled by the loss of species, have voiced concern a mass extinction is under way. Erwin, who is accustomed to assembling fossil evidence for such claims, tends to be more circumspect. "The problem as I see it is that a lot of people don't understand the difference between the kind of data that we use to examine mass extinctions in the past and the pattern of extinctions that have occurred up to this point." Most of the evidence in the fossil record of past extinctions tends to be things such as mollusks and snails—creatures that are common and geographically widespread. Many of the species that have recently gone extinct—*recent* meaning in the human era—are creatures specific to certain regions and tend not to show up in the fossil record.

The best example is the passenger pigeon. About 180 years ago there were 5 billion of these birds in North America. Now there are none. Is this evidence of mass extinction? It's hard to say. There's nothing to compare the disappearance of these birds to, because from the standpoint of a paleontologist they hardly existed in the first place. Only two fossils of passenger pigeons have ever been found—and one of them was discovered in San Diego, which is outside the creature's range. "It's not to downplay the damage we've already done to the planet," says Erwin. "If you compare apples to apples instead of apples to kumquats, a lot of the things we've lost so far are things that we simiply don't see in the fossil record."

Erwin thinks it's perfectly plausible that we're in a mass extinction event, but as a paleontologist, he cannot point to evidence either way. "Most of the people who claim that we're in the middle of the sixth or actually it would be the seventh mass extinction don't know anything about mass extinctions.

Because if that were true, the only thing you could recommend is going out and buying a case of really good scotch, because we're screwed."

But for our maudlin purpose, let's leave that debate aside for the moment. Let's assume the worst, that we are currently in a Holocene mass extinction, and think about what it looks like, and perhaps we can see where it might be headed.

IMAGINE THAT A race of extraterrestrial geologists lands on Earth 250 million years from now, exhumes the fossil record, and finds that a magnificent race of hominids once dominated the planet, the way anaerobic bacteria dominated the planet 2.4 billion years ago, and then some horrific event killed off nine of every ten species. All the higher creatures were wiped out—the humans, of course, but also the deer and the cats and the elephants and the wolves. A few rats were left, perhaps some snakes, lots of insects, and humorless ocean fauna. How would these alien geologists reconstruct the present era? When did it begin? How fast did it happen?

Ordinarily we think of an event as something that happens over a few hours, or perhaps a weekend, or a few weeks at the most. The Olympics is an event, an earthquake is an event, and so is a hurricane, a storm, a birthday party, a college reunion. Extinction events may not be much longer in duration. Scientists tend to talk about them as happening on geological time scales, but that's because they're reading the evidence from the geological record, which isn't usually very granular. Just because the geological record paints with a broad brush doesn't mean big changes didn't happen on a particularly bad weekend—as in the case of the dinosaurs—or over some other human time scale.

When we talk about the Holocene extinction event, what time scale do we use? Climate scientists tend to date the beginning of human impact on climate to the start of the industrial revolution round about 1850, and when they spin their climate models into the future, they'll often refer to how things may be in 2050, or at the end of the century.

Human activity, of course, amounts to more than simply carbon emissions. You could argue that our impact on the planet began one hundred thousand years ago when *Homo sapiens* migrated out of Africa and eventually replaced Neanderthals and *Homo erectus* as the dominant hominid species. Or you could argue that the start of agriculture about ten thousand years ago was the true beginning of our transformation.

Since it's an extinction event we're talking about, let's for argument's sake start the clock at the moment humans first started wiping out the woolly mammoths and mastodons and the big mammals of North America, about fourteen thousand years ago.

This was the tail end of the ice ages, a time when the planet's climate reeled between ice ages and warm periods and back again. Only a finite amount of water is on Earth, so whatever is held on land in the form of ice—glaciers—is not available to the oceans. During cold periods, glaciers formed, and the sea levels dropped.

The continents were pretty much where they are today, but the coastlines were different. In what today is Alaska, instead of archipelagoes stretching out toward Siberia, you had a finger of solid land. About 15,700 years ago, the climate was beginning to warm and the glaciers were beginning to retreat. The ice covering the land bridge had melted enough to allow humans to make the crossing, but enough glaciers still remained to keep sea levels low. Hunter-gatherers began to wander eastward, picking their way between the two main glaciers of Canada or along the moderate coast, perhaps in canoes and kayaks, down through British Columbia, Washington, Oregon, to California and beyond.

The migration was quick. People came eastward and southward, settling both northern and southern continents. These were the Clovis people, so called because of their spearheads first found in Clovis, New Mexico. The same style of spearheads is found throughout both continents, suggesting that the continents had been empty of human inhabitants before the land bridge opened them to the migrants.

To get an idea of what life was like when humans were relative newcomers, let's pick a spot on the migration route down the Pacific—the Bay Area around San Francisco. Since the sea level was lower, there was no San Francisco Bay at all, just a lush green valley, with the San Jaoquin and Sacramento Rivers draining through a narrow break now spanned by the Golden Gate Bridge. The shoreline extended in a plain miles beyond present-day San Francisco; what are now the Farallon Islands are distant hills. The plains were covered with sedges, with some trees here and there, and were teeming with life.[16]

The rivers attracted a host of Pleistocene mammals we wouldn't know today. The short-faced bear, for instance, stood thirteen or fourteen feet tall when on all fours, about 25 percent larger then the grizzly, making it the largest carnivore ever found in North America. It could run at forty miles per hour, and it would have had little trouble running down camels, llamas, horses, or antelope. The ground sloth was as tall as a giraffe and as bulky as an elephant. The Columbian mammoth, a slightly smaller creature than the woolly mammoth, would coat itself in mud and relieve the itch caused by the parasites that lived on its skin and, before the mud could dry, would rub itself against the rocks. Some of these rocks are still there, polished to a high luster in a manner that suggested it hadn't been from wind, rain, or water, but rather from repeated rubbing by the mammoths after their mud baths. [17]

The extinction of the Pleistocene mammals happened quickly. In all, North America lost thirty-five types of animals, mostly large mammals, and many more species. Six genera, or groups of species, became extinct in North America, and twenty-nine disappeared entirely from the planet. The short-faced bear died off, as did the American lion and the dire wolf. Small animals such as the short-faced skunk, the Aztlán rabbit, and the dwarf pronghorn, which was about the size of a golden retriever, were also soon gone.

Over the last few decades, a debate has raged in academia over to what degree the Pleistocene extinction had to do with human hunters. The region's climate was also changing, for reasons that had nothing to do with the hunters. The age of glaciers was ending, and changing habitat puts pressure on the big predators.

In recent years, however, the research has pointed toward humans as drivers of the big-mammal extinctions. Jacquelyn Gill of the University of Wisconsin at Madison analyzed the dung of large mammalian herbivores found in ancient lake-bed sediments. She could tell how many herbivores were alive at the time by measuring how much of the fungus *Sporormiella* the dung contained. When you have a lot of herbivores eating a lot of plants and producing a lot of excrement, you get a lot of fungus, which release a lot of spores into the air. Some of these spores land in the lakes, where they are trapped in sediments.

Gill also measured the pollen contained in the dung, which told her about local vegetation. Charcoal in the dung indicated how much smoke from forest fires the herbivores had inhaled. She found that the end days of the herbivores were times of great change. With fewer of the beasts around to munch on broad-leaved trees such as black ash, elm, and ironwood, these plants thrived and over took the landscape. The collapse of herbivores meant more woody debris for firestorms, which wracked the forests.[18 19]

Many of these creatures were fierce and powerful and large, but they knew no humans. To hunters who had recently arrived, they were easy prey. To be sure, bringing down a woolly mammoth would have had its challenges—you can imagine how big a band of hunters it would have required, each one hacking away with spears and axes—but the creatures may not have bothered to flee. One woolly mammoth could have gone a long way to supplying provisions for the long, cold winters. The first Americans may also have used kayaks fashioned from the skins of small mammals.

Over the years, the big animals died off—some suddenly, others over thousands of years. Mammoths declined quickly about eleven thousand years ago; indeed, archaelogical sites that follow the Clovis finds have no signs of mammoth remains. There's still some debate over the cause of the extinctions. "But if we start with the notion that it's human-induced, for the sake of argument, that's really the beginning of our transformation of the biota of the planet," says Erwin. "And it's accelerated in the last one hundred years. We're certainly transforming the planet in lots of ways."

THE PLEISTOCENE MAMMALS may have been the first species that humans dispatched, but they were not the last.

The island of Hawaii provides a useful microcosm of what's going on all over the planet, as biologist E. O. Wilson points out.[20] Before humans settled the islands, they contained as many as 145 species of birds found nowhere else. They had native eagles and long-legged owls. "On the ground a species of flightless ibis foraged alongside the moa nalo, goose-sized flightless birds with jaws vaguely resembling those of tortoises, Hawaii's version of the Mauritian dodo," writes Wilson. All told, Hawaii once held more than ten thousand different types of native plants and animals.

Then the Polynesians came and hunted the flightless birds to extinction. They brought the first pigs, which roamed the forests wreaking havoc on native trees, wallowing in mud and leaving depressions where puddles formed and mosquitoes bred, spreading disease. Still, when Captain James Cook discovered Hawaii in 1778, he noted bananas, breadfruit, and sugarcane covering the lowlands and foothills. But soon colonists came to clear forests and grasslands for agriculture. Other alien mammals came, too—mongooses and rats, goats and cattle.

Today, Hawaii is still beautiful, but it's a ghost of its former glory. Almost all of the native bird species are now extinct. Of the original 145 species of native birds, only 35 are left, and 24 of those are endangered, perhaps beyond saving. In place of the original diversity, Hawaii has now been infused with vegetation and animals imported from elsewhere. Of 1,935 flowering plant species on the islands, says Wilson, 902 originated elsewhere, and these dominate the landscape. It's the ecological equivalent of every town having a CVS and a Starbucks. What's happened in Hawaii has happened at a larger scale throughout the world.

How far has the Holocene extinction gone? It's hard to know, not least because nobody knows how many species existed in the first place. Between 1.5 million and 1.8 million species have been named, but 100 million or so more may not have been discovered. Based on the area of habitat lost, estimates put the number of extinctions at 140,000 species per year, but there is no figure that everyone agrees on.

Species loss doesn't necessarily spell the end of the world. The trick is figuring out how much is too much. And that's the subject we turn to in the next two chapters.

(The short answer is: less than you think.)

CHAPTER 3

Climate Change

MARTEN SCHEFFER IS SOMETHING of a contradiction. He comes from a long line of accomplished doctors and scientists. His great-grandfather Rudolf Scheffer was a renowned expert in tropical plants and director of the Dutch botanical gardens. On his mother's side, her great-grandfather, C. L. van der Burg, was a physician who spent a great deal of time in Indonesia when it was a Dutch colony. He wrote two fat books from his experiences—one on the nutrition and diet of Indonesians, and another on the colony's indigenous diseases. According to family lore, after dinner he would pour himself a glass of champagne, settle into a thick leather chair, and write, pausing once in a while to feed a drop of the bubbly to his pet gecko.

The generation of Scheffers who came of age in the 1960s rebelled against such stuffiness. They chose the arts over the sciences. One of Marten's uncles is a sculptor whose work adorns many public spaces in Holland. His aunt became a belly dancer. His parents were musicians; they played piano and flute and taught school, out in the countryside among the apple orchards on land that sits a few feet below the level of the Rhine. As a boy, Marten used to play on the dike that holds the quick-flowing waters at bay.

At an early age, Scheffer developed a special fondness for water. There was always some nearby, clean and clear and cool, to jump into. Ditches and canals crisscross the Dutch countryside, draining and directing the water from the lowlands up, seemingly against gravity, eventually to the sea. It's no wonder that a boy growing up in the lowlands has an intimate connection to water. "That's how it all starts, I suppose," he says.

When it came time to go off to study at the University of Utrecht, Marten briefly considered music. As a child of musicians, it's come naturally to him—he plays classical violin, and also guitar and mandolin. In the end, however, he rebelled against his rebellious parents. He chose biology.

Scheffer looks more like an artist or a musician than a scientist. He is balding on top; what hair he has is long and trails off Einstein-like in wisps. He keeps a short goatee and wears jeans everywhere. His mind is also prone to making leaps and connections that would seem to go against the grain of the plodding methods of science, with its rigid specialties and caution and leather chairs. When he speaks, he may start talking about the ecology of ponds and lakes, but the topic eventually jumps to mandolin playing, or dinosaurs, or climate change.

Upon graduating, Scheffer eschewed academia and instead went to work for an environmental agency of the Dutch government. He looked into what cities could do to accommodate breeding birds. After that he looked into the effects of habitat fragmentation on isolated populations of animals—some animals, such as deer and badgers, had difficulty crossing roads and canals. Eventually, though, he wound up where he belonged, after all those years of jumping into the ditches and catching frogs and bringing them home as pets: at the Institute for Water Management, or RIZA. His first job was to look into what to do about lakes and ponds and ditches that suddenly go from clear to turbid.

The phenomenon, known as eutrophication, was fairly well understood. Farmers and homeowners were using more and more fertilizer. During a rainstorm, it would run off into local ponds, disrupting the balance of microbes and vegetation and fish, allowing algae to grow. The water turned dark and the surface covered over with pond scum.

Scheffer's bosses at the institute were well aware of what was causing the problem but were puzzled as to what to do about it. Removing the flow of fertilizer didn't seem to help—the algae remained.

That's how the problem landed on Scheffer's desk. After the chemists were unable to solve the problem, Scheffer and some other biologists were asked to come up with a way to fix it.

Many young biologists find inspiration in Darwin, but Scheffer drew his from "The Lake as a Microcosm," an article published in 1887 by the American naturalist Stephen A. Forbes.[1] "It described the web of interactions among organisms as they struggled to survive in a pool of water left after a retreating flood," recalls Scheffer. The account reminded him of the ancient Greek notion of mikros kosmos—the world in a grain of sand. If you can understand how a pond works, thought the young Scheffer, you might be able to gain some insight into more complex systems, such as jungles or oceans or everything else on Earth, or Earth itself.

Scheffer couldn't have asked for a better assignment than the eutrophied lake. Mountain lakes are deep, and they form layers, which sets up complex interactions that depend on weather patterns and a host of other externalities. Oceans are more complex still, with their planetwide currents, the interaction with the atmosphere above, the influence of freshwater runoff from land, and so much else that is unknown and unknowable. But a pond is something you can fit onto a lab bench.

So that's what Scheffer did. He and his colleagues set up bottles and aquariums that contained a representative slice of pond life. This little microcosm was much easier to study than an entire lake. You could take clear water with a web of microbes that formed a stable micro-ecology, and add nutrients—simulating what happened when fertilizer ran off into the local ponds—turn this mini-pond turbid, and then try to turn it back again. "The nice thing about acquatic ecology is that some of those systems we can just study in a bottle," he says. "We can study them in a very well controlled way, in microcosms, miniature worlds—which are worlds in the sense that they contain everything, the nutrients, the light, the temperature, the grazers—everything is there in a bottle. This is a system with a potentially very large complexity—hundreds of species—but allows us to scale it down and study it." In that way it's possible to begin to understand how a larger lake behaves.

Scheffer got this assignment not only because he had a background in ponds and lakes, but also because he also knew a fair amount of math. In those days, few ecologists bothered with it much, especially the kind of math that interested Scheffer. He studied a specialty known as dynamical systems, which deals with things that tend to change suddenly and thus are difficult to predict. It is the mathematics of the tipping point—the moment at which a "system" that has been changing slowly and predictably will suddenly flip. The colloquial example is the straw that breaks that camel's back. You can also think of it as a ship that is stable until it tips too far in one direction and then capsizes.

Scheffer's particular camel was the shallow lake. Add some fertilizer, and nothing much happens. Add a little more, and still nothing happens.

Add a little more, and suddenly the water turns dark, the plants die, and algae take over.

Scheffer suspected that the ponds were behaving as many dynamical systems do—they were switching from one "state" to another. In one state, the lake is clear. In another, it's turbid.

It seems like an obvious statement, but there's a subtlety to it, and that subtlety makes the difference between success and failure. By removing the nutrient flow of fertilizer runoff, the institute was assuming that they could get back to a clear state by following the same path back. That's where they erred. If you look at a lake as a dynamical system with two different states (clear and turbid), it's easier to see that getting a lake to switch back, from turbid to clear, requires a different series of steps. It's not enough to take the same steps in reverse, it takes a new strategy.

It's a bit like rolling a rock off a cliff. You push and the rock moves a few feet. Push again with the same force, and the rock moves again by the same distance. When you reach the cliff, though, suddenly a single push has a very different outcome. The rock doesn't move a few feet—it tumbles thousands of feet into the valley below. You could say that the system—the system here being the rock, the cliff, and you—has reached a different state. The first state is the rock at the top of the cliff, the second state is the rock at the bottom.

Clearly, getting the rock back to the top of the cliff is a very different job from pushing it off in the first place.

So Scheffer set up his ponds in miniature and began running his experiments. He found that if the water is clear, and it has plants growing on the bottom, the vegetation runs the show—it keeps the water clear by taking up excess nutrients, keeping the phytoplankton, also known as algae or pond scum, to manageable levels. The plants don't do this themselves—they merely provide a hiding place for zooplankton, tiny animals that filter the water and help keep it clear. The relationship between the plants and the zooplankton is mutually beneficial: plants need clear water or else they can't get the sunlight they need for photosynthesis. If no light reaches the bottom of the pond, the plants can't grow.

When a lake gets euthrophied—when nutrients from fertilizer flow into the lake—the zooplankton can't keep up, and algae begins to grow like crazy on the surface, blocking sunlight. The plants die. The zooplankton die. Algae takes over. You get a cascading effect, and it all seems to happen at once, as though someone threw a switch.[2]

All of a sudden, the ecosystem that keeps the lake clear collapses. The clearing effect of the vegetation dwindles, and it's a downward spiral. The vegetation goes in a rapid die-off—a mass extinction in miniature. The algae, with little competition, takes over, species die off, the biodiversity of the lake goes down. Fewer plants and animals can live in the lake. It becomes a kind of desert.

Even if you try to reduce the nutrients, there are no plants—and thus no zooplankton to clear the water anymore. If you restock the lake bottom with plants, they die because the water is dark. It is difficult to get out of this turbid state—it's like trying to roll a rock back up a cliff.

Scheffer's eureka moment came when he realized that the key to flipping back to a clear state lay in the fish. "We realized that fish play a very important role in these ecosystems. When a lake goes turbid, the kinds of fish that thrive are the ones, like bream and carp, that tend to stir up sediment, making the water cloudy, or eat the zooplankton that would otherwise clear the water." So they hit upon the idea of removing most of the fish in a turbid lake for a time—a few months—so that the plants and microbes had time to return.

One of the advantages to working in a well-funded government agency, as opposed to academia, is that it's easier to take what you do in the lab and put it into practice. Once Scheffer and his colleagues had perfected the technique of flipping experimental ponds from clear to turbid back to clear, it was time to try it in the field.

To start, he and his colleagues chose a tiny pond, about ten meters by five meters, that had gone from healthy and clear to a turbid mess. A local fishery that stocked lakes for sport fishing agreed to take the fish from the pond and keep them in a holding pond. Scheffer and his colleagues drained the pond, removed the fish with nets, and dumped them into trucks for transport to the holding ponds. Then they filled the pond back up. In a few months the pond started to clear, and the vegetation returned. When it seemed as if things were firmly heading in the right direction, they went to the sport fishery, fetched the fish, and restocked the pond. It worked.

They repeated the steps with a bigger pond, one that had served as a swimming hole for kids, but had in recent years developed a thick layer of pond scum. Although it wasn't quite as popular as it had once been, kids were still using it. "It was so overgrown with algae that people would come out green," Scheffer says. He and his colleagues drained it, removed the fish, and voilà—the swimming hole returned to its previous clear state.

The next step was to try it on Lake Wolderwijd, a turbid lake that was too big to drain. Scheffer and his colleagues came up with a method of sectioning off part of the lake with a dike, and removing the fish in that section. The idea was that if they could free even part of the lake and get it to regenerate, they could gradually take over the entire lake. Once again, it worked.

It may have been the first time biologists had been able to restore a large lake ecosystem like that in such a dramatic way.

By that time, Scheffer had published so many articles in scholarly journals that a friend suggested he try to get himself a Ph.D. He made some inquiries and found that his old school in Utrecht would grant him a degree if he turned in a thesis and defended it. His thesis, "Minimal Models as Useful Tools for Ecologists," was an anthology of his previous work. It covered birds and badgers, lakes and ponds and ditches. The defense was given in a seventeenth-century building in a wood-paneled room with oil paintings of famous professors and deans of the university. "It was like the Swedish Academy of Sciences in Stockholm, where they award the Nobel Prize," he says.

Scheffer's great-grandfather Rudolf Scheffer, the botanist, had sat for his thesis defense in the same room, under the same paintings. But Scheffer wasn't nervous. "I was pretty carefree about it," he recalls. "It was all so ceremonial. Perhaps being from a rebellious family, I wasn't intimidated. I like science, but I don't think it should be intimidating."

Utrecht awarded him a Ph.D. in 1990. Scheffer has since moved on from the institute to Wageningen University and Research Centre, where he is head of the aquatic ecology department.

He's also moved on from lakes and ponds. He recently got a grant to work on "generic early warning systems for critical systems." This is a fancy way of saying that he's trying to figure out to what degree the world behaves like a pond, and how to tell if it's about to suddenly turn dark.

In recent years Scheffer and others have begun to explore the notion that a pond has much in common with the planet's climate, in the sense that it can change gradually, then all of a sudden flip from one state to another.[3] The notion is speculative, but it's gaining currency among scientists. If it turns out to be correct, it means that Earth's climate is, or could soon be, on a knife's edge, ready to fall into some other mode, causing sudden, perhaps catastrophic, changes. And once it flips, it could be next to impossible to get it to flip back.

The idea that climate behaves like a dynamical system addresses some of the key shortcomings of the conventional view of climate change—the view that looks at the planet as a whole, in terms of averages. Scheffer's dynamical systems approach, by contrast, would look at climate as a sum of many different parts, each with its own properties, all of them interdependent in ways that are hard to predict.

It would create a whole new universe of possible ways the world might end. And it would mean that the most alarming of climate alarmists may turn out to be understating how bad things could get, and how quickly.

THE SCIENTIST WHO has most clearly and forcefully sounded the alarm is James Hansen.

Hansen, a brilliant climate scientist at NASA's Goddard Institute for Space Studies, has voiced concern about runaway global warming and the possibility of Earth turning into a hothouse devoid of life, like our planetary neighbor Venus—covered in thick clouds with a surface temperature that could melt lead.

In August 2005, Hansen got an invitation to deliver a lecture in honor of Charles David Keeling, one of the first scientists who turned his attention to climate. Keeling had died in June, and his son Ralph had extended the invitation to Hansen.[4] Hansen had been doing research on climate change for decades. The son of a tenant farmer in Denison, Iowa, Hansen wound up staying on at the University of Iowa to get a Ph.D. in physics, working under the tutelage of James Van Allen, who discovered Earth's radiation belts, and for whom they are named. Hansen went to work for NASA and did research on the clouds of Venus. For a time, he helped develop an instrument that was launched onboard Pioneer Venus, a robotic probe.

Venus is only slightly smaller than Earth, but it is closer to the sun and its atmosphere is more than 95 percent carbon dioxide. It is very hot—460 degrees centigrade on average—and its surface pressure is 92 times that of Earth's.[5] Venus, sometimes called Earth's twin, is thought to have once had Earth-like oceans, but these have long since evaporated. Keeling's work brought home a worrying parallel between Venus and Earth.

Keeling was the first to measure atmospheric carbon precisely. He was so diligent in taking measurements and keeping records, from his lab near his home in California, that he reportedly missed the birth of his first child.[6] Keeling showed that the concentration of carbon in the atmosphere decreased every summer in the northern hemisphere because vegetation would take up CO_2 as it grew, then give it up again when it died and rotted, driving concentrations back up again in the autumn. When he plotted this rise and fall on a graph, he got a line that zigged and zagged each year with the changing seasons. This is now called the Keeling curve.[7] What Keeling pointed out—and what got Hansen worried—was that the zigzag line had shifted upward—the lows and the highs were getting higher and higher each year. It was the first empirical evidence that human activity was increasing carbon dioxide levels.

Carbon dioxide is a greenhouse gas, which means that the more of it there is in the atmosphere, the warmer temperatures are going to get. Carbon levels are rising, and temperatures are rising. The burning question is, What will the planet do in response? Will it get warmer only gradually? Or will warming accelerate?

This question and others bothered Hansen so much that he gave up on Venus and diverted his attention to Earth.

In person, Hansen doesn't seem like an activist. He's shy and speaks quietly. But he had publicly criticized President George W. Bush's refusal to sign on to the Kyoto Protocols in 2001, which would have committed the nation to reducing emissions of greenhouse gases. Since Hansen was on the outs with the Bush administration, the invitation to speak at Keeling's memorial gave him a platform. He prepared a talk called "Is There Still Time to Avoid Dangerous Anthropogenic Interference with Global Climate?" He described "multiple lines of evidence indicating that the Earth's climate is nearing, but has not passed, a tipping point, beyond which it will be impossible to avoid climate change with far-ranging consequences."

By then, Hansen had come to the view that the warming will accelerate over time—that it will run away. The warmer it gets, the more conditions promote further warming, and so forth.

Temperatures fluctuate from day to day and year to year, and it's difficult to discern long-term trends. But over long enough periods—years, decades, centuries—climate becomes more predictable, variations cancel each other out, and it is easier to make predictions. You may not be able to forecast the weather next week, but you can get a decent fix on how climate is going to change over the coming decades by how much carbon dioxide we're adding to the atmosphere.

Hansen's worry is that at either extreme, hot or cold, climate would start to run away. In a certain middle range, Earth tends to be relatively insensitive in its response to adding carbon dioxide to the atmosphere. At either extreme—either very hot or very cold—the trend tends to reinforce itself. Hansen calls the hotter trend the Venus syndrome.

There is precedent for Hansen's view that Earth can, if pushed, go to one extreme or the other. The "snowball Earth" event of roughly 600 million years ago would be one such piece of evidence, though at the moment it's more of a hypothesis—that is, not settled.[8] On every continent, including in the tropics, there is evidence of glacial deposits from about 850 million years ago, which suggests that Earth had at one time been entirely covered by ice from the poles to the equator. Since the presence of ice at the surface leads to colder and colder temperatures, the question arises as to how the planet got itself out of this state. That's where the Snowball Earth hypothesis comes in. It holds that volcanoes spewed carbon dioxide into the atmosphere until eventually the climate began to warm, and as the snow melted, exposing water and land, the warming trend increased. Together, evidence of widespread glaciations and an explanation for how Earth might have escaped its snowball state supported the premise that Earth could slip into an extreme temperature state.

Hansen was sufficiently alarmed by this scenario that he came down off his perch of scientific independence and took to the picket lines. He has advocated discontinuing the burning of coal altogether, and in 2009 he was arrested at a demonstration at a mine in West Virginia.[9] If we burn all the available coal and oil, the Venus effect is likely, and if we also burn other sources of fossil fuels, such as tar sands and tar shale, says Hansen, "I believe the Venus syndrome is a dead certainty."[10]

What if Hansen is right about the Venus effect, but wrong on the timing? What if Earth's climate is less like something that smoothes itself out on average and changes over a few hundred years, and more like one of Scheffer's ponds—ready to flip?

Both Hansen and Scheffer use the phrase *tipping point* in reference to climate, but they mean different things. Hansen believes that climate may reach a point of no return, beyond which nothing will be able to reverse drastic warming. The warming, though, could take decades or centuries to occur. Scheffer talks about sudden changes—flips that happen with great immediacy. He is not so much talking about a global average, but the behavior of dozens of regional climate systems linked together, dependent on one another to varying degrees, and some of them prone to flipping from one state to another. A pond may exist more or less independently of the large ecosystem around it, but regional climate systems are *inter*dependent. If one flips, it could send others flipping, like a line of dominoes.

This is catastrophic thinking, mind you, but it keeps some scientists up at night.

• • •

In the 1970s, the eminent British scientist James Lovelock formulated a theory of Gaia, which held that the Earth was a kind of super organism. It had a self-regulating quality that would keep everything within that narrow band that made life possible. If things got too warm or too cold—if sunlight varied, or volcanoes caused a fall in temperatures, and so forth—Gaia would eventually compensate. This was a comforting notion. It is also most certainly wrong, as Lovelock himself wrote later. "The climate centres around the world, which are the equivalent of the pathology lab of a hospital, have reported the Earth's physical condition, and the climate specialists see it as seriously ill, and soon to pass into a morbid fever that may last as long as 100,000 years. I have to tell you, as members of the Earth's family and an intimate part of it, that you and especially civilisation are in grave danger."[11]

With the betrayal of Gaia, the world has come to look less like a benevolent organism that will always right itself no matter what we humans do to it, and more like a sick patient. Lovelock bemoaned the audacity of thinking that we could engineer the planet—that we could become such a big presence in the planet's climate and biology without screwing things up. As a species we are no doubt having a systemic impact on Earth's systems. So it might be a good idea to step back and look at what kind of systems those are, and how they might respond. We're poking the creature in the cage, so let's have a look at the creature itself.

Climate scientists, as we've said, have tended to look at Earth as a vast whole in which fluctuations average themselves out, the better to see long-term trends emerge. Another way to look at the Earth is as a huge network of smaller but interrelated systems whose behavior, individually and collectively, is something of a mystery.

Climate, for instance, is not one thing that covers the globe. It is a sum of weather patterns and ocean circulations and storms and droughts and clouds. It starts on an astronomical level, with the Earth's orbit varying as it goes around the sun. The orbit itself tends to vary slightly in several ways; sometimes it traces more of a circle around the sun, sometimes more of an ellipse. This affects the relative amount of radiation falling on the Earth throughout the year. Right now, Earth gets about 6 percent more energy from the sun in January than it does in July. But when the orbit is at its most elliptical, that differential climbs to 20 or 30 percent. This eccentricity varies in one-hundred-thousand-year cycles. The tilt of the Earth on its axis tends to vary, too, every forty-one thousand years or so, and affects the contrast between the seasons and the ease with which glaciers can form. Earth also wobbles on its axis, much as a top wobbles as it begins to slow down, which imposes another twenty-three-thousand-year cycle.

Taken together, these three things add up to complicated variations, called the Milankovitch cycles, which affect the amount of the sun's radiation that reaches the surface. Oddly, the changes in the sun's radiation from the Milankovitch cycles tend to have bigger effects on climate than you'd expect just from the variation in sunlight. That means that other things going on in the planet's climate system amplify these effects.[12]

A big factor is ice. Ice on the ground is white, which reflects sunlight and heat energy back into space. This "positive feedback" reinforces whichever trend is already under way. More ice reflects more sunlight, which makes a cold Earth even colder. By the same token, when ice melts, it exposes the ground or the water beneath it, which tends to absorb more energy and turn it into heat, thus reinforcing a warm planet's tendency to get even warmer.

Greenhouse gases—carbon dioxide and methane—are another factor. In the past, periods of rapid warming have been accompanied by rises in concentrations of greenhouse gases. The chicken-or-egg question climate scientists ask is, Which comes first, the warming or the greenhouse gases? From analysis of ice-core data in the Antarctic in the past 240,000 years, it appears that the temperature rises tend to come first, and the greenhouse-gas buildup lags.[13] This is not comforting. It means that a rise in temperatures triggers the release of greenhouse gases from the oceans and the soil, which thereby promotes warming—another case of positive feedback.

Earth is a gigantic circulating engine. Underneath these planetary cycles and feedback loops you have regional weather systems, each with its own set of complicated dynamics. Sunlight falls heavily on the equator, heating the air and causing it to rise. When it cools, it sinks back to the ground, heats up again, and rises. If the air heats up over a rain forest, say, where the leaves of the trees promote the evaporation of water, loading the air up with moisture, it will tend to cause clouds and precipitation. If the air sits over a desert, it will remain dry. Imagine the equator as a line of heat, and air rising and drifting back down to the north and the south, where it begins another cycle. Meanwhile you've got the jet stream snaking its way from west to east around the globe.

The oceans are part of this picture, carrying heat from one end of the globe to another. From *The Fate of Greenland*:

> The conveyer is one of the ocean's great global current loops. It originates in the northernmost regions of the Atlantic Ocean where, during the winter, frigid air flowing off Canada and Greenland cools, and hence densifies, the salty waters carried into this region by the Gulf Stream. The result is that surface water becomes dense enough to sink into the abyss to form what is known by oceanographers as North Atlantic Deep Water. This water drifts southward down the length of the Atlantic. When it reaches the tip of Africa, it joins the ocean's mighty Mixmaster, a circular torrent driven round and round the Antarctic continent by the force of the southern ocean's westerly winds.
>
> Balancing the Atlantic's inflow to this torrent (and also the input of deep water generated along the margins of the Antarctic continent) are outflows into the abyssal Indian and Pacific Ocean. The ocean's global current loop is completed when these waters well up to the surface and flow back to the Atlantic. The part of the upwelling that occurs in the Pacific is largely channeled through the Indonesian Straits into the Indian Ocean and from there around the tip of Africa back into the Atlantic.[14]

Add up all these cycles and feedback loops, and you've got one complicated, unpredictable beast. Now we are approaching the kinds of things that scientists have a difficult time calculating with certainty. They are not *deterministic*—you can't extrapolate future behavior from the past, and you can't assume that if you add 10 percent or 100 percent more carbon to the air, temperatures will rise proportionally. Rather, climate systems have elements that behave according to the rules of dynamical systems. If you poke a sleeping tiger, it goes back to sleep. What will happen if you poke it one more time is anybody's guess.

AT RELATIVELY RECENT time scales—within the past hundred thousand years—we get a much more detailed view of climate, and it isn't pretty. We see whipsaws in temperature that would make your head spin. In recent years, scientists have begun to go back and piece together the climate record from ice cores in Greenland and other clues, and they've found that climate is something of a hot potato. Temperature goes up and it goes down; more disturbingly, it seems to jump all over the place, changing gradually for a while and then making sudden leaps. In the past one hundred thousand years or so, sudden warming of ten degrees centigrade or more in the span of a decade has happened perhaps a dozen times.[15]

About fifteen thousand years ago, round about the time our ancestors were hunting mastodons in Canada, temperatures in central Greenland jumped in a few years almost fifteen degrees centigrade, then fell back down about ten degrees. They hovered there for a century or two, then plummeted again. Then they hovered for a few centuries, and all of a sudden, within a decade, they jumped ten degrees and continued to rise, albeit at a slower pace, for another century.[16]

Ten degrees doesn't sound like much. Consider that ten degrees of average temperature is enough to distinguish two very different climates. It is the difference between Houston and Baltimore, or between Columbus, Ohio and Minneapolis, Minnesota.

Scientists know about these periods of rapid warming and cooling because they've drilled through the ice sheet in Greenland and Antarctica and pulled up ice cores, which contain clues to annual temperatures. Climatologists can piece together temperature to about one hundred thousand years back by reading these ice cores of Greenland. As the snow falls, it accumulates, and it never completely melts, ultimately forming glaciers. The snow and ice pile up at the center of Greenland, then flow slowly downhill, toward the ocean, calving off into the sea. It takes about one hundred thousand years for a snowflake that falls at the summit of Greenland to work its way down to the sea. By drilling the two or three kilometers from the surface to bedrock, and studying the layers of ice for clues as to Earth's climate, scientists have been able to piece together a pretty clear view of the planet's climate. And it's a roller coaster.[17]

The early settlers of Greenland lived through one such rapid climate switch.

In the early 980s, Erik the Red landed with a small crew on the coast. He found a region of steep fjords and verdant valleys, cut by cool water flowing from snowcapped mountains. He named the new country Greenland. He established a colony at Brattahalid, which means "steep slope," and another about four hundred miles to the north, in what is now called Western Settlement. The climate was cold and unforgiving in the far north, but these men came upon a rugged land that was unpopulated, and they decided to settle there and see how they might fare. In 986, twenty-four boatloads of settlers set out from Iceland to join them; fourteen arrived bringing four hundred people.

The new lands were not quite paradise, but they were pretty nice. In the green valleys, cool, blue water flowed from the glaciers many miles away. There were big mammals to hunt, with pelts for making clothing and shelter. Along the thin strip of moderate coast, the settlers found enough flat land to grow hay for the livestock. The fishing was plentiful—trout and salmon from the streams, cod from the sea. During the dark winter months—the sun sank below the horizon for months at a time—they salted fish and buried it in the tundra, uncovering it as they needed it and feasting in the dark. It was not a life of leisure, but it was probably as happy as a hardscrabble human life can get.

The settlers prospered, and their numbers grew. Ruins suggest perhaps five thousand people at the settlement's peak. They built monasteries with stained glass and bells, played board games, raised sheep and goats, and traded polar bear pelts and live falcons with Norwegians in exchange for timber and wine. If we had been there to watch, we may have thought they were on their way to establishing a culture, a city, perhaps eventually a nation. The settlements prospered for five hundred years—dozens and dozens of generations.

In those days life was hard, and nature loomed large in people's lives. Existence depended on a delicate balance of weather and food sources. One winter things got pretty rough. The temperatures typically dropped way below zero, but this winter they went further still, and the winds blew particularly hard. That wasn't so bad, or even particularly unexpected. But the next spring came almost a month later than usual. The settlers' supplies ran out, and, never far from the edge of survival, they began to starve.

Babies were stillborn. Elders died. The villagers dug shallow graves in the rugged terrain and piled up stones, as tradition called for. Eventually summer came, and things got better.

Perhaps it was bad luck. But the next winter came earlier than expected, and it came harshly. Because of the late spring, farmers hadn't had enough time to grow and harvest enough crops to last the winter. The fishermen's wives hadn't finished their salting. The snow came and the wind came and they huddled for warmth, battened down the town for the winter, hoping things would go well.

They didn't. The winter was harsh and long, the following summer brief. Parents began telling their children stories of lush summers and calm winters, of harvest feasts and winter-solstice feasts and partying and great cheer. The children listening to these stories grew up and began telling the same stories to their children.

And then one day everyone was gone.

Nobody knows quite how this happened. Perhaps a mammoth snowstorm killed the last survivors. Perhaps the hard weather won a battle of attrition, as farmland gave way to erosion. Slowly people died and others left. Perhaps within the living memory of the elders, a way of life, a worldview that had started with the optimism of settlement, simply ceased to exist.

The settlers of Greenland wouldn't have been the only ones affected by such swings. Half a world away, in the caves of Southeast Asia, the record of rainfall is written in the layers that have accumulated on stalagmites and stalactites as water drips down from the ground during rains. By analyzing these layers, scientists have found that these sudden climate shifts would affect the water cycle half a world away, causing the monsoons to switch on and off, plunging the region into periods of alternating drought and heavy rainfall.

The contemporaries of the Vikings were enjoying warmer climes than that of Greenland's. They were in the forests of Europe and along the rivers of Africa. But those places were not isolated from the temperature swings in Greenland. That much scientists can infer from other signs gathered in these places and pieced together with the ice cores from Greenland.

"Cold times in the north typically brought drought to Saharan Africa and India," writes climate scientist Richard Alley in *Scientific American*. "Two centuries of dryness about 1100 years ago apparently contributed to the end of classic Mayan civilization in Mexico and elsewhere in Central America. In modern times, the El Niño phenomenon and other anomalies in the North Pacific occasionally have steered weather patterns far enough to trigger surprise droughts, such as the one responsible for the US dust bowl in the 1930s."[18]

THIS IS ABOUT as far as we can get from the Gaia hypothesis—a world that protects its inhabitants like a mother protecting its offspring. The notion of Gaia may not have been correct, but it got people thinking about Earth as a kind of cell, with mechanisms to cope with the vicissitudes of life. Or perhaps as a pond, with interrelations among the creatures who live on it, the geological forces below their feet, and even the cosmos.

One of Lovelock's scientific offspring is Tim Lenton, a scientist at the University of East Anglia in England. Lenton is two generations removed—his mentors were mentored by Lovelock—but as a second-generation

Lovelockian, he has the advantage of youth and an emotional distance on the subject. Lenton is a climate scientist, which means he is well versed in the language of complex mathematics and computer models, but he also has considerable knowledge of ecosystems and biology. He has been working for the past few years on taking the kind of notions that Scheffer applies to ponds and applying them to the globe.

"We are looking quite hard at past data and observational data that can tell us something," says Lenton. "Classical case studies in which you've seen abrupt changes in climate data. For example, in the Geenland ice-core records, you're seeing climate jump. And the end of the Younger Dryas," about fifteen thousand years ago, "you get a striking climate change." So far, he says, nobody has found a big reason for such an abrupt change—no meteorite or volcano or other event that is an obvious cause. The absence of one of these big external forces for change suggests that perhaps something about the way these climate shifts occur simply makes them sudden. In other words, maybe they just flip.

Lenton is not nearly as interested in the past as in the future. He has tried to look for things that could possibly change suddenly and drastically in the future even though nothing obvious may trigger them. Let's now take a look at what he's come up with.

THE PHRASE GLOBAL warming is something of a misnomer. The warming of the planet is indeed a global phenomenon, but climate is inherently regional—by definition. Northern Europe tends to be gray and temperate, the American West is arid and sunny. Most of the work on models has to do with average changes, but that is a gross oversimplification of what climate change means as we experience it in our lives. Like politics, all climate is local.

Lenton has been studying tipping points in climate systems. Defining a climate flip, though, is not quite as easy as looking at a pond and seeing whether the water is dark and the plants are dead. A climate system is more abstract and more complicated than a pond, which makes it harder to identify tipping points, and harder to predict if and when they're going to happen. But Lenton has a short list of nine tipping points that we may see relatively soon.

Not included in those nine ways is Hansen's Venus syndrome. That isn't the kind of phenomenon Lenton is interested in describing. The Venus phenomenon, as Hansen sees it, is slow, something that happens over many years. It has a tipping point in the sense that small changes may trigger positive feedbacks, making the phenomenon irreversible after a certain moment. But Lenton studies things that change relatively quickly, without apparent reason. His tipping points, in other words, might be more worrisome than Hansen's Venus syndrome.

But Lenton is not a big believer in the Venus syndrome anyway. A big driver of the Venus syndrome is the release of methane from permafrost. As warming proceeds at the poles, particularly at the north pole, the permafrost begins to melt, and the organic matter—dead trees and leaves and peat bogs and so forth—become accessible to the microbes that cause decay, which releases carbon. Boreal forests hold 50 to 10 billion tons of carbon in the trees and the soil, says Lenton, while fossil fuel emissions run about 500 billion tons a year. Even if the forests gave up the carbon, in the form of methane, it would make things worse, but it would be unlikely to send the planet over the precipice. When it comes to methane, says Lenton, the change in temperature seems to proceed proportionally to the amount of methane that's been released—the opposite of the kind of flip that occurs in a dynamical system. A big belch of carbon from permafrost would do some damage, Lenton argues, but it would not by itself be disastrous. "There have been an awful lot of silly things said about runaway global warming," he says.

What's interesting about Lenton's tipping points are that they tend to be regional in scope. This is one of those places where the climate models get wonky. They seem to be all right for broad averages, but if you live in London you don't particularly care about the global average so much as what the weather is going to be that afternoon or weekend or winter. (Does anyone know what today's average global temperature is?) And even if you did care, climate models are notoriously bad about predicting how a two-degree centigrade temperature rise will affect the Indian monsoons or nor'easters in the United States, or whether the drought in the west will go away anytime soon.

Climate models, in fact, often don't do as good a job as they might in capturing abrupt changes. Zhengyu Liu, a climate-modeling expert at the University of Wisconsin at Madison, has what he believes to be the most sophisticated model yet made of the behavior of Greenland climate events in the past twenty thousand years, but the model doesn't take into account what might have happened when the flow of freshwater from Greenland glaciers on the Atlantic Ocean stopped, which may have been a key driver in the temperature fluctuations. When Liu runs his program, he has to insert the sudden change in freshwater by hand, so to speak.

Lenton characterizes his tipping points in terms of how quickly the transition from one state to another is likely to take, once a tipping point is reached.[19] Let's start with the quickest-acting one first—the Indian summer monsoon.

Each year, the sun shines down on the dark surface of the Indian Ocean, and moist, warm air rises and forms clouds. This rising heat and the moisture form a powerful weather system, a natural pump that pulls up water and moves it in vast quantities hundreds of miles to the mainland. The monsoon deposits rainfall on thousands of square miles of farmland. About a billion people, most of them poor, depend for their daily bread on crops that depend in turn on the reliability and regularity of the Indian monsoons.

India is a rapidly developing country with hundreds of millions of citizens who want to move into the middle class, drive cars and cool their homes with air-conditioning. It is also a country of poor people, many who still rely on burning agricultural waste to heat their homes and cook their suppers.

Smoke from household fires has been a big source of pollution in the subcontinent, and it could disrupt the monsoons, too. The soot from these fires and from automobiles and buses in the ever more crowded cities rises into the atmosphere and drifts out over the Indian Ocean, changing the atmospheric dynamics upon which the monsoons depend. Aerosols (soot) keep much of the sun's energy from reaching the surface, which means the monsoon doesn't get going with the same force and takes longer to gather up a head of steam. Less rain makes it to crops.

At the same time, the buildup of greenhouse gases, coming mainly from developed countries in the northern hemisphere, has a very different effect on the Indian summer monsoons: it acts to make them stronger.

These two opposite influences makes the fate of the monsoon difficult to predict and subject to instability. A small influence—a bit more carbon dioxide in the atmosphere, and a bit more brown haze—could have an outsize effect.

Lenton believes once the monsoons begin to flip from one state to another, the change could happen in one year. What happens then? It's not a question that Lenton can answer with certainty. But he foresees two possibilities.

One is that the monsoons grow in force and intensity, but come less frequently. We have already seen hints of this in the newspapers. In the last few years rains have grown erratic and less frequent, but when they do come, they tend to dump an enormous amount of water, and in places where they wouldn't normally do so. This is almost as bad for farmers as drought, since the rain falls on parched ground with extra force, and much of it runs off without soaking into the ground, and it causes damage to boot by washing away soil and plants. The flooding that devastated Pakistan in 2011 is a case in point.

If this trend continued and strengthened in intensity, it would be bad news for the two thirds of the Indian workforce that depends on farming. It would be nasty for the Indian economy—agriculture accounts for 25 percent of GDP. A permanently erratic and harsh monsoon would depress crop yields, increase erosion on farms, and cause a rise in global food prices as India is forced to import more food.

The other possibility is even worse: the monsoons could shut down entirely. This would be an unmitigated catastrophe. A sudden stopping of monsoon rain, which accounts for 80 percent of rainfall in India, could throw a billion people into danger of starvation. It would change the Indian landscape, wiping out native species of plants and animals, force farms into bankruptcy, and exacerbate water shortages that are already creating conflict. The Indian government would almost certainly be unable to cope with a disaster of such proportions. Refugees by the hundreds of millions would stream into big cities such as Mumbai and Bangalore, looking for some hope of survival. It would create a humanitarian crisis of unprecedented proportions.

On the other hand, a flip in the Indian monsoons could land somewhere between these extremes. The Himalayan mountains will continue to force warm, moist air coming off the ocean to rise, creating some rainfall, but there's no telling how much rain there would be, or where under a new climate it would fall. When the monsoon system flips, the past will be a poor guide to the future, and the break will not happen gradually. That is the nature of tipping points. "It would not be a smooth transition," says Lenton. "It would be a jump to different mode."

Lenton foresees a similar danger of sudden change in the West African monsoon, which gathers warmth and moisture from the Atlantic equatorial regions during the winter, rises up the Atlantic coast, and deposits rainfall in the Sahel region of Africa, which includes Niger, Burkina Faso, and Mali. Much like the Indian monsoon, the African monsoon seems to be balanced on a knife's edge between conflicting forces. Greenhouse gases are increasing the amount of moisture available to the monsoon, while soot from pollution that originates in the northern hemisphere is weakening the monsoon. There are dozens of other complicating factors. The difference in sea-surface temperatures in the Atlantic on either side of the equator, which tend to be lopsided and prone to unpredictable fluctuations, could push the monsoon in either direction, toward becoming drier or wetter.

According to Lenton's estimates, a flip in the West African monsoon would take about a decade. Whichever way it goes, the change will be abrupt. It's hard to tell what a shift in the monsoon would mean for Africa. A collapse might shut down a major cause of rainfall, making things drier throughout the region. Or it could make for wetter conditions by allowing weather from the west to blow in with the jet stream.

It's easy to see what drier conditions would mean to the region. The countries of the Sahel are already struggling to find enough water, as periods of prolonged drought continue to lengthen. "Most of the countryside is an immense sweep of infertile windswept scrubland," wrote Scott Johnson for *Newsweek* in 2008. "Prolonged droughts and flooding have been problems here for as long as anyone cares to remember." In the past half century, say scientists, rainfall across the Sahel has dropped by 25 percent, and 120,000 acres of arable land is turning into desert each year. Despite these conditions, the Nigerois keep having babies—the population has quadrupled in that time to 13 million people. By the end of the century, the UN predicts, Niger will have a population of more than 100 million.

Drier conditions would mean that the drought that took hold of the "triangle of death"—an area of eastern Africa that includes Somalia, Ethiopia, and northern Kenya—in the summer of 2011 would become the new normal. The drought, the worst in fifty years, brought acute malnutrition to much of the population of Ethiopia and Kenya.

It's difficult to exaggerate the severity of the humanitarian crisis. Somalia has lacked a central government since 1991. Food prices tripled in the Somali capital between 2010 and 2011, largely because of the failure of local crops and poor distribution. Deforestation has had a huge impact. Trees are gone, and grazing land is drying up. In insurgent-held areas, 3.5 million people were in danger of starving to death during the drought.

The other possibility for the West African monsoon, of course, is that it causes increased rainfall to Africa. This would be a good thing in many ways. It could conceivably turn the Sahara desert into a rain forest.

The Sahara desert, a vast area north of the Sahel, receives less than an inch of rainfall a year in parts. It is the biggest desert in the world, except for Antarctica. It is a land of giant sand dunes and blinding dust devils. But it wasn't always so. Dig beneath that sand and you find fossil remains of hippos and other big game, similar to what's found now on the African savannas far to the south. Scientists have analyzed core samples of the ocean floor that contains dust from the Sahara over time and have found evidence that for thousands of years it was a lush rain forest. Then, about five thousand years ago, the rains stopped, and the region turned into a desert. The cause: a flip in weather patterns.[20]

What's interesting about the Saraha flip is that if you had lived there before this event you would never have been able to predict it. Things seemed to be evolving in a slow, gradual fashion, but then suddenly changed dramatically. The flip was likely caused by a subtle variation in the orbit of Earth, but even though Earth's orbit changed slowly and predictably from year to year, this time it suddenly brought rapid change—the lush vegetation died, the animals died, and the sands took over. It's possible that what we're looking at for the region is a return to ancient weather patterns.

There's no guarantee, though, that a flip that caused the West African monsoon to deposit more rain on the continent would return the Sahara to its prior state. It's just one of many possible outcomes.

Tipping point number three in Lenton's list is the sea ice of the north pole. For years the ice has been thinning and retreating more and more during the summer. Soon it may disappear completely during the summer months. We may already have reached this tipping point—a transition to a new state in which the north pole is ice-free during summer months is already at hand. Eventually the north pole may flip and be free of ice year-round. The knock-on effects of such a transition would be huge—they would cause marked increase of warming at the north pole, since open water absorbs more of the sun's energy than ice-covered seas.

The effect of a year-round ice-free north pole would be like heating Greenland on a skillet.

The fourth tipping point is Greenland's glaciers, which hold enough water to cause sea levels to rise by more than twenty feet. It takes a while for that much ice to melt, of course. Currently, IPCC (Intergovernmental Panel on Climate Change) projections say it will take on the order of a thousand years. Scientists currently don't have a good handle on how such a big hunk of ice melts. For plenty of reasons it could happen much more quickly—recent observations suggest that the melting has not only exceeded what models predict, but has also begun to accelerate. A marked retreat of ice in coastal areas has led to an infusion of ocean water, which is relatively warm and promotes melting.

All this leads Lenton to conclude that the Greenland ice sheets could make a transition to an alternate state in three hundred years, rather than a thousand or more. If you think that sounds like a long time, consider that the melting could happen in fits and starts.

A relatively quick melting of Greenland would have a knock-on effect on the ocean currents that run up the Atlantic, bringing warmth to northern Europe and Scandinavia. The current in question is the Atlantic thermohaline circulation. This current has two modes—the existing one, and an alternate mode that would stop short of the British Isles by a few hundred miles or so. A switch to the alternative mode could plunge much of Europe back into an ice age. Scientists were getting nervous about this possibility a few years ago, until further research suggested that any switch in current is a long way off—perhaps a thousand years off.

Yet there's little certainty. For one thing, an accelerated melting of Greenland would throw more freshwater on the northern Atlantic than the calculations have taken into account. A sudden influx of freshwater would change conditions from what the original predictions were based on, and they could accelerate a flip in the thermohaline circulation.

"The canary in the coal mine is the Arctic losing its summer sea-ice cover," says Lenton. "I am really worried about the Greenland ice sheet. It's already losing mass and shrinking."

The question is, what will the Greenland ice sheet do from here on out? It is already shrinking faster than computer models predict, which suggests scientists don't understand something fundamental. If it is approaching a tipping point where it will slip from one state (covering the entire continent) to another, what will that other state be?

The less pessimistic scenario is that Greenland flips so that the ice sheets melt partially, leaving the bulk of the ice on land, where it won't contribute to sea-level rise. The ice sheets are presently like a hand that rests largely on land but is sticking its fingers into the water. Most of what's melted so far is thought to be the fingers. The best-case scenario would be if the fingers melt but the glaciers above them somehow stabilize. The continent would flip into a different state, but with lots of ice still in the highlands, with the coasts greener. That would lead to something like a one- or two-meter rise in sea levels over a period of time that's hard to pin down—probably a few hundred years. This scenario does not count sea-level rise that comes from other effects, such as the expansion of the oceans due to warming, or to additional meltwater from the south pole.[21]

The worst-case scenario for Greenland, however, is that the massive ice sheets that lie above the fjords are melting as well and will continue to melt. In the language of dynamical systems, Greenland would flip into a completely ice-free state. This would cause massive rises in sea level—on the order of six or seven meters—just from Greenland alone. Even if this took three hundred years to happen, "it would be an absolute disaster," says Lenton, "a real game changer." At such a rate of sea-level rise, it would become more and more difficult to protect coastlines. Low-lying areas would have to be abandoned. That includes cities such as New York, Los Angeles, San Francisco, London, Tokyo, and Hong Kong, not to mention the entire state of Florida and vast swaths of Indochina.

Tipping point number six—the west Antarctic ice sheet—is even scarier. It has enough ice on it to raise sea levels by about eighty meters. The ice is melting, but slowly—most worst-case scenarios give the ice centuries to melt. But there are some niggling doubts about whether the West Antarctic Ice Sheet could calve into the sea more quickly than expected, as the glaciers contract. If that happened, it would push sea levels up by five meters in as short a time as a century. Although most experts consider this unlikely, Lenton thinks the sheet could flip in as little time as three hundred years, which is three times faster than most models predict. Lenton does not know whether or not this transition has begun, or when it might begin.

Researchers in Germany and Holland recently spun out scenarios of what a five-meter rise in the next century would mean.[22] The seas would inundate many low-lying areas. Much of Florida would disappear, large areas around the North Sea, including Holland, and major deltas in Asia. The Nile River would rise and flood the delta, putting parts of the city of Alexandria underwater. The waters of the Mekong River would rise and consume its delta. Many rivers that run through major cities would back up due to the rising sea and necessitate the building of barriers or the abandonment of low-lying areas.

One of the most vulnerable regions would be the Dutch lowlands of Scheffer's childhood, with its system of dikes and dunes and sluices, the networks of canals and pumps. The Dutch have currently designed their flood protection on the assumption that sea level will rise 0.8 meters over the next century—which is well above the predictions of the IPCC. Should the Antarctic ice sheet begin a precipitous collapse, most likely the rises would be ignored for several years, or perhaps a decade, before the public could be persuaded to do something about the problem, and the government acted to bolster its defenses against water. Corporations might start taking action sooner rather than later and move their operations to the eastern, higher parts of the Netherlands, or to other nations entirely. People living in these abandoned parts of the nation would feel the impact first in the closing of plants and office buildings and the loss of jobs. Local economies would feel the effects. Restaurants would close, schools would lose students, stores would shutter their doors, rents would plummet, and more people would go on the dole.

The general abandonment of low-lying areas would create feelings of unease and vulnerability. A terrorist attack on a dike, say, could push people into a panic. Dutch society, which at the moment is vibrant and resilient, might begin to seem frayed and vulnerable.

The big Dutch cities would not be immune from the decline. Rotterdam, one of the great shipping capitals of the world, would lose business to other cities if shippers began to have problems managing the flow of goods or protecting ships from the weather. Amsterdam, already the "Venice of the North," would grow even more watery. Buildings that were built alongside canals and waterways would be prone to flooding. The city would get a reputation for being moldy and wet and unpleasant, and tourists would stop coming. Property damage would skyrocket. Mold remediation companies would thrive.

It would take time for the government to act because governments are inherently slow, even the small Dutch government. The Dutch are so liberal that you might expect that leaders would be unwilling to give up on the relatively poorer low-lying areas of the north and west, but they would be powerless to prevent the flight of commercial firms and the increasingly mobile intellectual workforce that goes along with them. Indecision would set in, and steps taken to support evacuees would be too little, too late. The result would be growing economic disruption, poverty, and social chaos.

The city of London has also come under researchers' purview. London years ago built the Thames Barrier, a series of gates that can be raised at high tide to prevent a storm surge from raising the level of the Thames upstream and to protect the floodplain upon which much of London stands. Engineers built the barrier after years of flood damage, and Queen Elizabeth inaugurated it in 1984. Since then engineers have been upgrading it to handle a one-meter sea-level rise in coming years, in line with conventional climate-change projections. But an Antarctic event would make those plans wholly inadequate, and this would become apparent over a few decades.

A big storm might breach the Thames Barrier. That would hasten plans to upgrade it, but sooner or later another storm would breach the upgrade. London would experience its Hurricane Katrina. Waters would rise and inundate the Houses of Parliament and Victoria, and they'd lap up against the walls of Buckingham Palace. (The queen would, of course, have retreated to Balmoral by then.) The damage would be counted in the tens of billions of pounds.

The Dutch and German study group projected that a rapid rise in sea levels would outrace the ability of engineers to protect the London floodplain. Eventually, the UK would have to abandon large swaths of central London and relocate to higher ground. Or the city would have to rebuild itself as a twenty-first-century Venice, with canals and gondoliers with Cockney accents.

The damage to the developed world of sea-level rise pales in comparison to what would happen to the developing world, where people are already living on the edge, and where governments don't have the resources to cope even with the slow sea-level rise that conventional climate models predict. Rising waters in Bangladesh and the Mekong Delta would create tens of millions of climate refugees. At least 400 million people live within five meters of high tide, and if current population trends continue, this figure will double by the end of the century.

WATER AND ICE aren't the only worries. The Amazon rain forest, the seventh of Lenton's tipping points, is also in jeopardy.

For decades now, loggers and ranchers have been clearing forest, and the Brazilian government has been carving it up with roads. As a result, the forest there has become prone to dieback, which means the area of lush, green, and wet tropical forest has begun to shrink.

Rain forests are always pretty wet, but they have dry seasons, and those dry seasons turn out to be a limiting factor on the survival of flora and fauna. As loggers reduce the number of trees that produce moisture to feed the gathering rains, the drier the dry seasons get, and the longer they last. Lately dry seasons in the Amazon have gotten more severe and have put a crimp on the survival of many of the trees that form the forest canopy, which is the backbone of the rain-forest ecosystem.

As the dry season continues to lengthen, the flora draw more and more water from the soil, which eventually begins to dry out. The trees get stressed and begin to die. There's more fodder on the forest floor for wildfires. This is not hypothetical; it's already begun to happen. We saw this during the estimated twelve thousand wildfires that occurred in the Amazon during the drought of 2010.[23] As the forest loses more and more trees, it loses its ability to feed the weather patterns with warm, moist air.

The tipping point comes when we can no longer stop this trend, and the rain forest trees reach a critical point where the entire region turns into something else entirely—something other than rain forest. What would that be? Maybe a drier, deciduous forest, with a whole different set of flora. Or perhaps the rain forest will turn into an open savanna, with mainly grasslands and sparse trees. Whichever state the Amazon flips into, it would be drier.

Scientists have known from research done in the 1980s and 1990s that the Amazon has had a big effect not only on local climate but also on circulation patterns that reach far and wide. The Amazon is basically a big spot of wet tropics. Knock out the trees and lose that moist air, and the regional circulation pattern changes as well.

Air circulation is complex, but one basic fact of moisture is simple to understand: take moisture out of one part of circulation, and you increase it somewhere else. The Amazon supplies big masses of warm, moist air, moving upwards. Take that out of the map and replace it with a dry land surface and you're weakening the current flow pattern.

When you think about having to map out all the moving air on the planet in such a way, with land masses getting in the way and altering where the flow goes, and add in a jet stream, and storm systems, you begin to get a feeling for how interconnected the whole thing is.

Something similar to what's happening along the Amazon is happening in the boreal forests of Canada—the eighth of Lenton's tipping points. Those vast forests have in the distant past been prone to strikingly abrupt transitions—they made sudden shifts between tundra and forests as the ice ages advanced and retreated. Instead of tropical trees under threat, in the boreal forests you've got pine and deciduous trees under stress from increasingly dry weather. Trees are further badgered by new pests, most notably the pine beetle, which has more success in breeding as climate warms.[24] Warmer temperatures have encouraged the microbes that cause forest decay. As trees die, there's more tinder for forest fires, which have been on the rise. Partially because of these fires, boreal forests have recently gone from being net absorbers of atmospheric carbon (via photosynthesis) to a source (via smoke and animal effluvia).

What many scientists fear is a complete die-off of the forests and, in their place, new grasslands or prairie. The northern forests don't contribute moisture to the weather as the Amazon does, but a die-off would add considerably to the amount of carbon that gets released into the atmosphere. In northern areas, ground that had largely been frozen for most of the year—permafrost—is beginning to soften year-round. That makes 50 to 100 billion tons of organic matter available to microbes, which munch away at dead leaves and trees and release methane, a potent greenhouse gas. (As noted earlier, this would not cause runaway warming, but rather would contribute one of many regional climate flips.)

The basic weather patterns that we've grown used to on weather maps are also subject to rapid change. Among them is what's called El Niño–Southern Oscillation—the ninth and last of Lenton's tipping points.

El Niño involves movement of a blob of warm water on the west side of the Pacific Ocean toward the east, bringing with it moist warm air. When this warm water cools and circulates back westward, El Niño comes to an end and La Niña begins. These two patterns alternate roughly every five years.

From observations, scientists have begun to see a more erratic trade-off between these two patterns. They fret that the weather patterns could flip to some different state—though what that state would be is hard to say. Lenton thinks that a change could lead to more frequent switching off between the patterns, and perhaps more pronounced effects—in other words, El Niño would be warmer and moister. That would have a detrimental effect on the Amazon, says Lenton, exacerbating trends that already threaten to destroy the rain forest.

As these different weather systems begin to wobble like a top that's about to fall over, the jet stream that blows from west to east around the globe could begin to drift northward or southward. The jet stream isn't a tipping point per se—there's no way scientists see it suddenly shutting down or behaving in some extreme way—but there's no saying that these other weather systems wouldn't affect it in ways that cause problems for those on the ground. Floods in Pakistan, extreme heat in Russia that has led to unprecedented wildfires, drought in the western United States and Australia, an unusually dry winter in the northern hemisphere in 2011–2012—all of these trends could be worsened by a shifting jet stream.

The real nightmare scenario is when all these changes begin to reinforce one another. The Arctic loses its summer sea ice, causing Greenland's ice to melt and encouraging the boreal forests to change as well. The freshwater runoff changes the thermohaline dynamics and affects the jet stream. The El Niño–Southern Oscillation and the Amazon interact in such a way as to reinforce one another, perhaps affecting the monsoon in India and Africa. "It wouldn't be such a silly thing to say that if you meddle with one, you might affect the other," says Lenton. "Which direction the causality would go is not always obvious. We know it's connected, we know it's nonlinear, we know they somehow couple together. When you see one change, you see changes in the other."

"Then we start talking about domino dynamics," says Lenton. "The worse case would be that kind of scenario in which you tip one thing and that encourages the tipping of another. You get these cascading effects."

The ultimate disaster would, of course, be a mass extinction event. As we've seen in the previous chapter, humans are about as likely to survive one of those as a dinosaur standing in the Yucatán 65 million years ago, wondering what that funny bright thing was in the night sky. With Lenton's cascading climate tipping points, we begin to see how such an extinction event could conceivably happen without a meteorite impact.

Indeed, sudden flips that caused mass extinction may already have happened.

The extinction event that occurred 2.4 billion years ago, you'll recall—the Great Oxygenation Event, in which algae produced oxygen that killed off much of the planet's anaerobic bacteria—happened on a time scale that scientists have only been able to pin down to within a few million years. That sounds like a slow-motion event, but it wasn't necessarily. Algae, you'll recall, did no harm as long as there was enough iron to absorb the oxygen the algae produced; once all the iron had rusted, the oxygen built up in the atmosphere and death came to the anaerobic bugs. But here's an interesting twist. In 2006, scientists discovered an inexplicable lag—on the order of 300 million years—between the moment the iron had completely rusted and the moment everything died.

It's difficult to know precisely what happened when in "deep time"—events that took place billions of years ago. But Lenton has a theory for what caused this lag. In those early days, the planet teetered between two states: one in which oxygen levels were fairly low (less than 1 percent of the atmosphere), and another in which oxygen levels were high (about 20 percent, as they are today). The buildup of oxygen may have happened slowly, over millions of years, but the switch could have come suddenly, like a circuit breaker going off, delivering the anaerobic bacteria a sudden death blow. The event happened so quickly that the bacteria didn't have time to evolve and adapt to the new world.[25]

How quickly did this shift occur? Lenton's best guess is two hundred thousand years.

That may still sound like a long time, but consider that we are looking at an event that took place 2.4 billion years ago. Guessing more precisely is like hitting a dartboard from the moon. A few hundred thousand years is getting awfully close to a human timescale. We are in no danger of flipping to some oxygen-poor state, of course. The point is that flips can occur—may already have occurred—on the scale of human experience, without warning, and with disastrous consequences.

WHAT DO WE do about all this? One answer is to study these tipping points and figure out a way of anticipating them—perhaps with a view to stopping them from occurring.

As Lenton explains it, when you measure things such as temperature and wind and circulation, you get a variation in readings that are analogous to a rock's being rolled toward a cliff. Over time, if you do your measurements diligently, you might begin to tell if something odd is happening—you can see the curve of the cliff, before the rock rolls over the edge.

But this is all a budding science, and anyway it's not clear that tipping points could be discerned with time enough to make the necessary changes to head them off. The drawback is that once climate shifts to a new state, it may be too late.

One of the best recent examples is the weather in Australia, which seems to have taken a more or less permanent turn for the worse. The nine-year drought that began in 2001 and ended in 2010 with massive flooding could be the near-term future of places such as the American Southwest.

Nobody paid the drought much mind when it first hit, in 2001, the year Australia pulled out of the Kyoto treaty. Drought has always been a big part of life in Australia. In many parts of the world, average yearly rainfall actually tells you something about what to expect. In Australia, knowing average rainfall tells you almost nothing about what's going to happen next year.

The next year, however, the drought persisted, and in 2002, for the first time since anyone could remember, the Murray River failed to reach the sea.

The Murray River basin is a vast area, the size of France and Spain combined, and it makes up the agricultural heart of Australia. Although the southwest is known as the wheat belt, the Murray basin accounts a wide variety of farms and 40 percent of the continent's farm economy and 85 percent of the nation's irrigated land.[26] The water that flows down the Murray River is siphoned off by farms along the way, so that by the time it reaches the sea, even in good times it's lost much of its force.

The Great Drought of the Aughts wasn't a drought in the conventional sense of the word. You think of a drought as a lack of rainfall over a stretch of country year upon year, and you tend to think of it as being a fairly localized phenomenon. But this isn't quite what happened in Australia during these years. The drought hit just about every facet of the continent, from the wheatlands of the southwest to the rain forest in the north to the Murray-Darling Basin in the southeast. Sometimes one area got relief that turned out to be temporary. Indeed, farmers in the basin would hope that the worst was over, only to be hit again the following year.

Rice and wheat yields fluctuated wildly during the drought. Rice is a particularly water-intensive crop, and one wonders why Australia, a dry land, would grow it. (Farmers there argue that they've managed to get productivity, measured as yield per acre, unmatched anywhere else in the world.) At the start of the decade rice production had reached a peak harvest of 1.6 million metric tons. By the last year of the drought it had fallen to a mere 18,000 metric tons—a drop of 99 percent. Wheat production fell by 50 percent.[27] These falls had a big impact on world markets. Australia's wheat exports account for 15 percent of world markets, and its rice exports account for 25 percent.

"I have a great graph of rice production in Australia" during this decade, says Peter Gleick, president of the Pacific Institute, a water policy think tank. "It goes to zero. They had to stop growing rice. Water rights went out the window. They had to scramble."

The Murray-Darling Basin was hard hit. Water turned salty and acidic, killing gum trees for a stretch of fifteen hundred kilometers along the river. The Menindee Lakes, up the basin, and the Coorong Wetland, near the mouth, have both deteriorated due to lack of freshwater to keep salt water at bay. Shorebirds, pelicans, black swans, and fairy terns died in droves. Kangaroos, in a desperate search for food, started showing up in city streets and parks. Several species of eucalyptus trees died out, putting the koala bear at risk.

Farms failed during the drought years in Australia at a rate that is rarely seen in the developed world. The drought caused an estimated drop in agricultural production of $7.4 billion (Australian dollars) and a loss of seventy thousand jobs in 2002–03 alone, says the Australian Bureau of Statistics. "What we have just been through is more than twice as bad as the worst drought previously recorded, which was in 1902," said a grazier in central Queensland. In the millenium drought, he said, "We have had trees dying here. Those trees were here when Captain Cook sailed up the coast [in 1770], so when you have got brigalow trees dying, it is a drought—there's no doubt about it."[28]

Toward the end of the drought, things got so dry that it turned Australia into some kind of hell. Australia has always had wildfires. Aborigines used to burn the tall, dry grass to clear it for farming—a practice known as firestick farming. During the drought wildfires became more frequent and widespread. In 2009, things got so dry that starting in March, four hundred fires burned four thousand homes and buildings across 450 square kilometers of Victoria. Then there were the dust storms. In September of that year, a cloud of dust five hundred kilometers by a thousand kilometers formed in the outback and began to move east toward Sydney. The skies grew black. Air traffic was disrupted.[29] It was the worst dust storm since the 1940s, reported the Bureau of Meteorology.

Through most of Australia's history, government has tried to encourage people to settle the rural lands and make their way as farmers and tame the hardscrabble outback. They encouraged irrigation, and in times of drought, which are as certain in the outback as death and taxes, the government has always been there to put farms on life support. But over the twentieth century industry grew and farming shrank in relative economic importance. By the 1990s, the government began to inch toward reform of these measures. Perhaps, some people wondered, it might be a good idea to begin charging for irrigation rights, to encourage a more efficient use of water, which was becoming precious. But farmers dragged their feet, and for years it seemed as though the water policies put forth in the eighties and nineties would never take hold, and for a while it seemed they never would.

Once the drought got under way, however, opposition to the water reforms crumbled. In 2007, Prime Minister John Howard proposed drastic cuts in water use, and his successor, Kevin Rudd, signed them into law in

2008. Sandra Postel, the economist, called it "perhaps the boldest water reform of this type ever proposed."[30]

Australia is drier than any other continent and is highly variable in rainfall. Its climate is influenced to a great extent by the El Niño Southern Oscillation, which causes droughts and wet periods. It gets pounded by the tropical cyclones, heat waves, bush fires, and frosts often linked to the vicissitudes of this weather system. High air pressure zones sit over the continent, keeping Australia's skies clear and free of rain, sometimes for weeks at a time. Many temperate zones don't have rainy and wet seasons, but tend to have spring showers and thunderstorms in summer. In Australia's temperate south, there's less predictability than in most places. Studies at the University of New South Wales have recently shown a high correlation between warm sea-surface temperatures in the Indian Ocean and drought in southeastern Australia (the Murray Basin region), which suggests that another mechanism, poorly understood, may have an even more pronounced effect than El Niño.[31]

Nobody knows when rain is going to fall in Australia, nor where it's going to fall, nor how much. And nobody can say what effect climate change is going to have on average rainfall. The climate models tend to predict constant average rainfall over Australia. But this isn't something to bet the mortgage on. Rainfall has decreased 15 percent since the middle of the twentieth century, according to a report by CSIRO (Commonwealth Scientific and Industrial Research Organization), and temperatures have risen about as much as you'd expect from the global coverage. The lack of rain and drier air cause water to evaporate from the soils and decrease the amount of rainfall that runs off, because the soil is drier and soaks it up. Overall, water availability in the Murray Basin will decline by 35 to 50 percent, says CSIRO, and the amount of water that reaches the mouth of the Murray could drop by 70 percent.

This means that Australia can expect more of the same drought, which is not a good prognosis.

Australia's drought had an effect on the rest of the world, not least because it added to food price rises. But that would be insignificant compared to what would happen if, say, China, with its 1.3 billion people, had to absorb a similar disaster. In China, farming and irrigation are already being pushed to the brink. If China's agricultural output were to take a big hit, it would have substantial consequences for the rest of the world.

For a foretaste we need look no further back than the drought of 2011. The people of Hubei have a folk song that goes "Oh, waters of Honghu Lake, wave after wave. The fisherman live ever better, year after year . . ." Nobody was singing much that spring. The year was shaping up to be the worst drought in the region in fifty years. From January to April, the province of Hubei was down 40 percent in rainfall. In the adjacent province of Jiangxi, Poyang Lake, China's biggest freshwater lake, had shrunk to a fraction of its full size, leaving fishing boats stranded on grassland that once was lake bed.[32]

China has been growing at a breakneck pace, with close to 10 percent GDP growth a year for the past decade or so, hardly pausing during the 2008 economic collapse that has so vexed the developed world. It has not been happening cost-free, though. Growth requires electrical power. China has had to expand its supply of power rapidly, which it has been doing by building coal and nuclear plants and hydroelectric plants. To feed the beast China has been building dams, an estimated fifty thousand along the Yangtze River alone, including the world's biggest, the Three Gorges Dam.

Water is inextricably bound up in all this growth and development. China has tapped its rivers to irrigate its farms. The Yangtze, Asia's biggest river, winds its way through 4,000 miles of the Chinese countryside and empties into the East China Sea; about 80 million people live along the Yangtze River Delta. To the north, in Hubei province, the water table is dropping. Only a few years ago, freshwater was 20 or 30 meters below the surface. Recently, villages have had to dig 120 to 200 meters. This has taken a toll on wetlands, which are drying up. It has also caused conflict in the past three decades. Village militias clashed over water rights in the 1970s. In the early 1990s, the villages of Huanglongkou and Qianyu were at war with one another over the construction of new canals. They exchanged mortar fire for the better part of a decade, culminating in a New Year's Eve battle in 1999 that reportedly resulted in one hundred deaths.[33]

Beijing, a city of 20 million people, is a thirsty city, and to satisfy its water needs it's been drawing more from surrounding Hebei province. Indeed, Beijing recently took control of the Juma, the last major river flowing through the province. Downstream, the waters have pretty much petered out, forcing 3 million Hebei residents to rely on groundwater, for which they must drill deeper and deeper.

To fix its growing water problem, China, a nation run by engineers, is of course trying to engineer a solution. It is building huge waterways to divert water from the south, where it's been plentiful, to the north. But in 2011, drought hit the south as well.

Peter Gleick, a water expert at the Pacific Institute and certified genius (he was made a MacArthur fellow in 2003), publishes every two years an assessment of the world's water situation. He doesn't like to talk about worst-case scenarios. "The worst case is a nightmare," he says. "I don't see the point in alarming people. And anyway, what does 'worst case' even mean?"

Despite his reticence, Gleick is not sanguine on China. "There is absolutely a danger of crop failure," he says. "They are extremely vulnerable to extreme events. Because they have a large population, food production is on the edge of satisfactory. They are vulnerable to droughts. The big population centers are in the wrong places, from the standpoint of hydrology. Would a failure cause famine? I don't know. China today is not the same as China thirty years ago. They can outbid anyone on world markets for food if they have to."

A crop failure in China might not bring famine to the Chinese, but it would mean famine for someone.

CHAPTER 4

Ecosystems

HUMANS ARE A HARDY SPECIES, and we've thrived in part because we are so adaptable to circumstances. We are capable of eating just about anything and everything. In a recent study of the Sanak Island ecosystem, in Alaska, over the past six thousand years or so, hunter-gatherers have a unique place vis-à-vis other species. Jennifer Dunne, an ecologist at the Santa Fe Institute in New Mexico, has been working up a mathematical model of the Sanak ecosystem, and it's apparent from her work why humans have done so well.

The Sanak archipelago stretches westward from Alaska across the Pacific Ocean toward Siberia. The settlers there came over when the glaciers of the last ice age covered most of the northern hemisphere, and stayed, living off the bounty of the intertidal pools, and the seals and sea lions and fish off shore, the salmon of the freshwater streams, and bird eggs of the tundra.

A quick survey of hunter-gatherers on Sanak makes it clear that they were both super-generalists and super-omnivores. As super-generalists, they ate a wider variety of different things than most other species. As super-omnivores, they ate more things at every level of the food web, from seaweed at the bottom to sea lions at the top, and just about everything in between. Dunne and her colleagues are finding that in a hunter-gatherer society, this quality works to the advantage of an ecosystem. Dunne has been reconstructing the seafood menu of the typical Sanak hunter-gatherer of six thousand years ago, and more generally on what all the creatures on that menu ate. Much of this information is already in the scientific literature; what researchers don't find there they look for in the field. They'll sweep a net through a stream to catch aquatic insects and dissect them to see what's in their guts. But data on the role of humans in the ecosystem had not been gathered before. Dunne and the other Sanak researchers went to a wide variety of sources to glean what they could about the fish and plants that Sanak hunter-gatherers ate. They talked to archaeologists, ethnographers, and present-day Aleuts who live in the region. They dug into ancient trash heaps of discarded shells and fish bones and excrement, preserved over the millennia, for clues as to what the Sanaks ate. They took inventory of all the species that inhabit the archipelagos now and what they know existed a few thousand years ago. They found that the food system in Sanak has been remarkably stable for thousands of years, which is surprising considering how radically the rest of the world has changed.[1]

Here's a Sanak food chain: At the bottom are phytoplankton, or algae, which drift around in the water, soaking up the sun and turning it, via photosynthesis, into energy. Shrimp eat the phytoplankton, only to be eaten in turn by the great sculpin, an odd bottom-dwelling fish. Steller's sea lions in turn eat the great sculpin. And humans—or at least the hunting, gathering variety—ate the sea lions, and probably cut big slabs of fat and cured them. Sea lions were also important for hides, which were used to make kayaks for hunting marine mammals and for fishing. That kind of food chain is simple enough, but things are of course much more complicated than that—there are thousands of food chains at Sanak, intersecting and doubling back on one another—and that complexity is just the thing that Dunne thrives on. The Sanak intertidal food chain consists of 176 different kinds of plants and animals connected to one another by 1,117 feeding links, and the researchers are adding more species and links as their research progresses. What they are describing is not so much a food chain as a complex food web. Dunne believes that she and her colleagues are the first to build a detailed and comprehensive food web that includes humans.

Dunne is one of a relatively new breed of ecologist who has a numbers sense. She's an expert in the mathematical relationships that describe networks—the same mathematics that underlies social networks such as Facebook and the phone network, but also networks of plants and animals that eat each other. You can look at food networks as a simple hierarchy, with plankton at the bottom and humans at the top, but those simple food webs don't describe the real world very well. Humans, for instance, feed at all levels. We eat predators at the top, plants at the bottom (think seaweed), and even fungus (mushrooms). We eat herbivores (cows) and the plants they eat (corn and grass).

More realistic food webs contain a great deal of complexity. You would include not only plants and animals but also nonliving organic matter—bits of dead tissue floating around in the sea, fish scales or pieces of skin, feces, bits of plants. Just as raccoons in the suburbs feed off garbage in Dumpsters, many animals scavenge rather than kill and many things don't kill to eat. Herbivorous insects, for instance, just take bites out of plants without killing them. Some animals—probably more than you'd think—are cannibals, at least when times get tough or their own species becomes too predominant.

Dunne has gone through the inventory of the food sources at Sanak, and who eats what, and represents those relationships in a computer as a bunch of nodes interconnected with lines. So the shrimp would be one node in this network. It would have a line going down to the plankton, but also to zooplankton, detritus, and worms. And it would have lines going upward to the things that prey on it: crabs, sculpin, and sea otters. Like humans, sea otters also eat most anything, but they also have to watch out for sharks, and humans.

For Dunne, the properties of these networks look like a symphony of numbers. She is a bit like Tank in *The Matrix*, seeing patterns in what looks to the untrained eye like a random stream of data. But each property of the Sanak Island ecosystem can be represented numerically, changing seamlessly from a series of instinctive interactions to a maze of numbers, and back again.

Dunne begins with simple observations, but, by applying a few mathematical techniques, gets critical insights. Most prey animals eat about six different kinds of species. Humans are the creatures who feed on the most sources of food—forty-seven—which comes to 27 percent of all the existing sources. Their biggest rival for the top spot on the food web on Sanak was the sea star, a many-armed creature, which feeds on forty-one different intertidal species.

Humans have a couple of other interesting qualities in the Sanak network, the way Dunne analyzes it. For one thing, you can't really say that humans are at the top of the food web because they eat on all levels of it. For each species in the web, Dunne calculated its average distance from the primary producers—the algae and the plantkton that get their energy from the sun—at the bottom of the web. (Since each creature is connected to the bottom of the web by more than one route, she averaged the species out.) Phytoplankton, for instance, would get a 1, while the periwinkle snail, which lives in the intertidal, would get a 2, and so forth. The average for the entire web turns out to be 2.34, which means that the average organism in this food web is less than one and a half "steps" away from the bottom. Sanak hunter-gatherers got a score of 2.82, which put them 35th from the top of the food web, behind the sea anemones, worms, whelk fish, sea lion, and sea otters.

Another way to slice it is to measure the average path length between species on the web. The path length between a predator and its prey—say, the shrimp and the phytoplankton— is 1. This may sound meaningless, but to a network expert such as Dunne, it gives some indication of the impact any one species may have on other species in the web. The shorter the path length between any two species, the quicker an overabundance or extinction of one will affect the other. In the Sanak food web, the mean path length is 2.4, with any individual species going from 1.59 to 3.31—the smaller the number, the shorter the species' path to other species, and thus the greatest effect it will have on others. The species group with the shortest path lengths are zooplankton (1.67) and detritus (1.59)—not a species per se, but rather the very small stuff that floats around or settles to the sea floor and serves as breakfast for many species. These two things establish a foundation for the rest of the food web. Detritus provides food for scavengers, phytoplankton is a primary provider of energy, zooplankton is eaten by a great many species. The Sanak hunter-gatherers, at 1.82, have the fifth-shortest path length. Humans, in other words, have the biggest potential impact on the Sanak food web of any species save some kinds of plankton and garbage.

This is unsurprising to anybody who has looked around at the impact of humans on the globe. Seven billion people have a big impact on the planet, there's no doubt. But what's interesting is that the Sanak hunter-gatherers did not destroy the island. Quite the opposite. They were key to keeping things stable. In the past six thousand years, the fossil and archaeological records suggest that the ecosystem of the island has stayed remarkably consistent. Although the balance between different species may have shifted, the inventory of species on the island has hardly changed.

The hunter-gatherers may have been largely responsible for this balance. Dunne is quick to point out that this would be an interesting hypothesis to test, but the researchers haven't yet done so. There is no obvious way, she says, to disentangle the importance of one species from the others in a complex food web. In any case, it's possible to say with certainty that the Sanak hunter-gatherers did not act in a way to upset the balance of the ecosystem.

The reason has to do with what Dunne calls "prey-switching." When the crabs are plentiful, they're at the top of the hunter-gatherer menu—it's crabs for breakfast, lunch, and dinner. But when the crab population drops off, it's time to try something else—whatever happens to be plentiful at the time.

> Nonhuman predators do this. They eat twenty different things, but they prefer two or three of them, because it's easier to catch, they have a high caloric content etc. Maybe because they are really focusing on a preferred item, they drive the population down somewhat, or maybe it goes down for other reasons and becomes harder to catch. You have to put more investment in the thing that's harder to catch and suddenly it's not preferred anymore. So the predator thinks, "Well, I can't really get at that so I'll get at this other thing." Prey switching is very natural, it is sort of bound to happen, in a way.
>
> We know that the humans in this system, the Aleut, did prey switch. When the salmon run came in, they would go drop everything and go and hang out by the creek and capture salmon. The salmon run would eventually end. Maybe it was stormy out, so they would go and do a bunch of harvesting from the intertidal zone. Maybe they had preferred sea urchins, but then they ran out of sea urchins so they switched to something else in the intertidal that was easier to get. Then maybe the storms lifted and they were able to go out in their boats and hunt marine mammals. But then it gets stormy again, and it's so stormy they have to stay on land, so they gather bird eggs. They're constantly switching depending on what's available, what the conditions are, how many calories they're getting, how easy they are to gather or hunt. So there are all these trade-offs, and it eventually gets them to switch.

But those were the hunter-gatherers who settled in Sanak. Most of their brethren went onward, however, to the great American continents, following the Bering land bridge. They would have found a very different scene—and they interacted with the fauna in ways that were much different from those who settled in Sanak. They hunted many animals to extinction.

We seem to have taken our cues from the first Americans rather than the Sanak hunter-gatherers.

What happened since then is an oft-told story, so we'll move quickly through it. We invented agriculture. We settled down. We had time to develop culture, and towns and cities, and eventually the megalopoles sprang up. Humans thrived, and over most of the next ten millennia, the population grew slowly. Then it started to grow quickly, and then very, very quickly over the past hundred years or so, despite numerous predictions of our own demise, all of which turned out to be wrong, of course.

The way our food economy works now is diametrically opposed to the way the Sanak hunter-gatherers' food economy worked. When a source of food becomes scarce, such as caviar or truffles, we tend not to switch to another more plentiful food item. Instead, we bid up the price and redouble our efforts to produce or harvest more. Rather than make use of a variety of food types, we specialize on only a few, because that's what it takes to achieve economies of scale, drive down the price, and feed more people. Our modern economy engages in the opposite of prey switching.

Not long ago, cod was so plentiful it was called chicken of the sea, and it was considered a bounty that would never end. But several cod fisheries have collapsed and others are on the verge of collapse. The cod fishery off Newfoundland peaked in 1968 at 810,000 tons, and then went into a freefall. Canada banned cod fishing in the region in 1992. The ban has yet to be lifted, and the fishery has not yet recovered.[2] Now cod is something of a delicacy. Some scientists think that we have been fishing down the food web—substituting smaller fish in place of the big ones that have been fished out—but this is more like a systematic depletion of resources than prey switching, which conserves resources.

Consider, too, the bluefin tuna. A bluefin has a body temperature of about eighty degrees, which makes it equally at home in the frigid waters of the arctic and in the warmer tropics. But the fish's great range makes it difficult to restrict fishing; a boat can always head to other waters for its prey. Predation pressure on bluefin in one area exerts pressure on bluefin in another. The Mediterranean bluefin is almost completely gone, due to overfishing, mainly from big ships that use long lines with multiple, baited hooks to catch the fish. The price of a bluefin has soared in recent years—a half-ton fish can fetch more than $50,000 in Tokyo.[3]

The disappearance of fish such as the bluefin is not just a problem of putting food on the table. The loss of a single predator species like that can have a knock-on effect. In the oceans, nine of every ten predators have disappeared, which has a far-reaching and potentially irreversible effect on ocean ecosystems.[4,5] Since it's harder to study the oceans than the forests and the savannas, scientists have a less thorough understanding of just how these effects play out. Predator species such as shark and tuna have been in decline, and there've been ripple effects. The decline of the blacktip shark off the coast of North Carolina, for instance, led to the proliferation of the cownose ray, the shark's prey. The rays have in turn led to the collapse of local bay scallop populations, destroying a fishery that had supplied food and supported livelihoods for more than a century.[6]

What goes for ocean ecosystems goes for ecosystems on land, too. The loss of wolves in Yellowstone led to a rise in the number of elk, who ate the cottonwoods and the willows, which in turns affects the insects that live and feed in those trees. It also affects the birds that live in the trees and beavers, which damn up the streams and thus change the ecology for fish downstream.[7] Erosion from the loss of trees can also change the landscape dramatically. All this from the loss of one species at the top of the food web.

The notion that ecosystems are on shaky ground is not going to come as a great revelation to anyone who keeps up with current events. The signs of decline are in plain sight. Bat populations in the Northeast are being decimated by White-Nose Syndrome, caused by a fungus, which kills them in droves in their caves.[8] Frogs in many areas of the western United States and even in the rain forests of Costa Rica and elsewhere are disappearing due to fungal infections that don't have a clear cause.[9] Bird migrations are changing as climate warms. Invasive species are running rampant in all parts of the world, taking up niches left by native species in their flight from changing climate.

What's really scary is that when ecosystems begin to get shaky, it could put our food supply at risk. It may be easy to forget that the crops that are harvested to give us our cereal and our potato chips and our Doritos are part of a global ecosystem, but they are, and they are subject to the same prospect of collapse that any ecosystem is. In fact, food crops may be more vulnerable than many ecosystems. At the very least, a problem with food crops would have a much greater impact on the fate of the species than problems with wolves or sharks.

To understand just how devastating a collapse of food crops could be, it's important to consider how we came to be so dependent on them in the first place. One reason the doomsayers turned out to be wrong about the pending destruction of the human race was because of the work of one man: Norman Borlaug.

Borlaug grew up on a small farm in Iowa. He became a plant pathologist, specializing in the study of the diseases of plants. As a young researcher Borlaug was invited to Mexico by the Rockefeller Foundation to see if he could do something about the country's wheat crop. The wheat was suffering from rust, a condition caused by a fungus that turns these golden plants into black, shriveled things with stems that collapse. Borlaug realized that Mexican farmers needed a high-yielding strain of wheat, amenable to fertilization, that would resist rust and thrive on irrigated farms. He and other researchers began crossbreeding thousands of strains of wheat to come up with something that would work.[10]

Borlaug took to this task with the single-mindedness that only the truly passionate can muster. He felt keenly that the fate of millions of people in poor countries throughout the world hung on his success or failure.

Keep in mind that Borlaug and his colleagues were crossbreeding plants the old-fashioned way: by growing them, cross-pollinating them, and looking closely at all the varieties they could produce, then trying again. These were the 1940s, decades before genetic techniques would be invented.

It took many years of trial and error, but Borlaug and his colleagues. came up with a strain of wheat that produced an abundant head, the most productive part of the plant in terms of food, and that didn't succumb to fungus. However, it couldn't survive until harvest. The stalk was too thin and the head too heavy, so the crop would fall over in the fields and die before maturing.

Borlaug took his new strain and went back to the drawing board. Eventually, he crossed his creation with a Japanese dwarf wheat strain and produced a miniature version with a thick stalk that could grow to maturity without toppling over. This new strain of wheat solved Mexico's problem, and within a few years farmers throughout the developing world began planting it. He had started what later became known as the Green Revolution, which helped create the conditions for the explosion in the world population in recent decades. In 1970, he won a Nobel for the achievement.[11]

Borlaug likely saved the lives of countless people by his actions. But the Green Revolution has also put us out on something of a limb. It has led to a vast global agriculture industry that is overdependent on only a few strains of four major crops—wheat, soy, rice, and corn.

The U.S. agricultural industry, for one, relies on a narrow genetic base of crop species. In 1990, six varieties of corn accounted for 46 percent of the U.S. crop, nine varieties of wheat made up 50 percent of that crop, and two types of peas accounted for all but 4 percent of the pea crop.[12] A similar system of monocultures is taking hold throughout the rest of the world, as more and more Asian countries ramp up their farming to Western standards of scale. More than half the world's potato fields are now taken up by a single species, Russet Burbanks, which McDonald's prefers for its fries.

That's a problem because it makes our agriculture industry prone to disease or disruption. One killer disease could wipe out vast swaths of our food supply. A billion people on the planet already live on the brink of starvation. Imagine what would happen if another billion or so were thrown into that category because suddenly the corn crops throughout the world collapsed from some kind of disease for which there was no ready cure. Imagine the financial disruption that would cause in developed countries.

Such a disruption may already have started, a decade ago, in Uganda.[13]

In 1998, farmers began to notice that their wheat plants, just before harvest, went from a healthy golden color to an ugly black; the grains shriveled and the stems broke. The cause: stem rust.

What was so perplexing to the scientists at the Kalengyere Research Station in October 1998, when they diagnosed the problem, was that for decades almost all commercial wheat crops have been varieties that contain one gene or other that confers resistance to stem rust. The crops being decimated in Uganda contained the gene Sr31, which supposedly made the crops invulnerable. The significance of this finding was not lost on the scientists. They realized that they were dealing with a strain of fungus that had found, through evolution, a way of thriving despite Sr31. This strain has put 90 percent of the global wheat supply at risk.[14]

Like many fungi, the Ug99 strain (named for the location and year it was discovered) produces tiny spores that easily become wind-borne and drift for hundreds of miles before settling onto another host. In this way, the fungus has since spread throughout Africa, causing wheat failures throughout the continent. While the fungus drifts, it also mutates. Scientists have identified seven different varieties, and evolution will no doubt keep producing more. In Kenya, a mutant form resists not just the Sr31 gene but also Sr24, another fungus-resisting gene, and has affected at least 80 percent of that nation's crops. Other varieties have surfaced in Ethiopia and South Africa. In 2006, fungus spores reached Sudan, and from there they crossed the Red Sea into Yemen and have been found as far as Iran.

From there, it won't take long for the spores to hop to Pakistan and India, which grow a fifth of the world's wheat crops. After that, according to calculations, the spores are expected to continue east on the winds, reaching China as soon as 2013.

Scientists have been scrambling to find new ways of reducing the vulnerability of wheat crops to stem rust. They have screened thousands of wheat varieties for those that have an inherent resistance. But they are running out of time. It takes at least five years to develop a rust-resistant variety. If the spores make it to India and China and deliver the kind of destructive power they showed in Kenya, when they devastated three quarters of the crops, the effect on the people of those nations, and on the global economy, could be significant.

The run-up in food prices we saw in the two years before the 2008 economic collapse will be mild compared to what will happen if stem rust hits Asia. The shortage of wheat will drive up the prices of rice, corn, and soy as well, affecting livestock producers, who have to buy grains to feed their cows, pigs, and chickens. In recent years, since the rise of bird flu, China has made great progress in improving the safety of livestock farms. A rise in grain prices would set that effort back a decade, increasing the chance that disease would spread among the farms. Livestock producers under the gun to slash costs to keep up with rising grain prices would start cutting corners.

If China were to be hit with a shortage of grain, it would divert grain production to its domestic needs. It has already been lining up deals with producers in Africa for arable land to cultivate, and these grains would wind up being diverted to China as well. The shortage of grain would bring hundreds of millions of people to the brink of starvation.

NO SINGLE FUNGUS is going to bring down civilization or cause the extinction of *Homo sapiens*. But taken together, they could be edging the world toward a state in which just a small additional shock could push it over the brink. Apocalypse may not be the most likely outcome. Such a fate would probably require a perfect storm of factors—the worst scenarios of climate and weather and ecological disruptions. But the list of factors that could contribute to such a perfect storm are rising. Colony collapse disorder has led to a mysterious decline in honeybees, upon which one trillion dollars of the world's agriculture depends. The banana crop, another monoculture, is already beset by yet another fungus. Superweeds are showing a resistance to Roundup, a leading herbicide, threatening to set back agriculture yields. The list goes on.

This book is really about the what-ifs. So, what if many of these annoying problems turned into major problems, and at the same time a few other things happened to put additional stress on the human population itself—things such as a killer flu, which we looked at in the first chapter. That's what we'll get into now: a few of the things that could push an already troubled world over the edge.

CHAPTER 5

Synthetic Biology

THE TECHNOLOGY OF DOOMSDAY HAS a long tradition with the country's military planners, going back to the days of the Manhattan Project. Planning for the next war, however, is different from planning for wars that won't be fought for decades. For that really long-term stuff, you need ideas—freewheeling ones that spring from well-informed minds.

One of the Pentagon's most productive methods of corralling the advice it needs about futuristic technologies it should be paying attention to is to hold quasi-public meetings of the smartest intellects, without requiring the participants to go underground and talk only in secrecy. The idea is that the threats under discussion are sufficiently far out into the future—sufficiently close to the science-fiction end of the spectrum, rather than the next big thing on the battlefield—that the risk of tipping our hand to the bad guys is minimal.

The JASONS are such a group. For many years, the JASONS, a few dozen of the nation's brightest scientists, have assembled once a year or so to discuss the latest wrinkle in nuclear weapons and other technologies of military significance.[1]

In the 1990s, the life sciences were in the national spotlight. The vast research apparatus of the National Institutes of Health and its affiliated universities and research organizations had begun the gargantuan job of painstakingly decoding a complete DNA sequence of a single human—a sequence of 3 billion base pairs, or letters.

This prospect was enough to give military planners pause.

What the Pentagon wanted to know was, when scientists succeeded in publishing the human genome—a map essentially of the biochemical possibilities at the heart of each and every human cell—what would this mean for the nation's armies? In the days of the Cold War, the Soviets avoided distributing completely accurate maps of Moscow; they didn't want any potential invading armies to know how to get around too easily. If the genome is a sort of road map of all the biochemistry of the essential characteristics of human life, then an accurate genome is a kind of map for invading pathogens, i.e., biological weapons, synthetic diseases, and so forth. Potentially, the human genome project could provide a blueprint for a bioterrorist, a map of detailed targets to hit. And it wouldn't take anything as cumbersome as missiles or uranium 235 to carry it out.

This all seemed urgent enough before biologist Craig Venter came along and galvanized the human genome effort. Venter, a brash genius, announced that he had techniques that would allow him to shortcut the methods of the NIH's effort to map the genome. Almost overnight he turned the slow march of progress into a mad dash to the finish line.

Military planners were well aware that they wouldn't be able to keep the first human genome under wraps, that it would be published for all to see. In the mid-1990s, it seemed that in a few more years, anybody in the world who wanted a road map of human biology would have access to it with a few clicks of the mouse. What would this mean for national security? What would the bad guys be able to do with all this information?

To find out, the Pentagon thinkers turned to the JASONS. And the JASONS turned to Steven Block.

Block was ideal for the job. The JASONS had always chosen people who had a capacity to think across many different disciplines. Block, a Stanford university professor, was interested in the intersection of physics and biology—particularly the behavior of DNA and other molecules in the cell's nucleus. Being well versed in physics, he was able to talk the language of much of the rest of the group, which at the time was dominated by physicists, a relic from the days when the biggest threat to humankind was the ability to split the atom.

Block is a shaggy, gregarious sort, and he likes to talk—in particular, he likes to talk about where science might go, and he's not afraid of speculating. This is a relatively rare thing, especially when it comes to biology and technology. Many scientists are circumspect—they'll talk about the potential of biotechnology to heal illness and create jobs, but they dummy up when it comes to the potential for misuse. This is entirely understandable.

There's no percentage in scaring people away from the pursuit of knowledge when that's your livelihood and you believe it's generally a good thing. Block is less inhibited. Some things he won't talk about, of course, for fear of giving a blueprint to would-be bioterrorists. In the summer of 1997 at the JASON meeting in La Jolla, California, Block gave them an earful.

The JASONS have high-level government security clearances, which means they became privy to classified, or at least sensitive, military information. They're not strictly prohibited from talking about the subjects they deliberate on, but they aren't in the habit of it. Yet Block did not fit this mold.

Block worried about all the biological weapons that were within reach. He worried about anthrax—a common bacterium found in cow pastures that could cause massive respiratory failure. He worried about plague and tularemia and brucellosis. He recalled the bizarre food-poisoning episode in September 1984 in Oregon, when 750 people fell ill (no one died) and two years later a follower of Bhagwan Sri Rajneeesh confessed that the group had released salmonella bacteria in salad bars.[2] And he recalled that Aum Shinrikyo, the Japanese cult that released sarin nerve gas in a Tokyo subway in March 1995, killing twelve people and sending five thousand to the hospital, had bought a five-hundred-thousand-acre sheep farm in Australia to carry out tests of botulinum toxin and anthrax.[3]

These kinds of terrorist opportunities weren't what most concerned Block, however. The emerging methods of biotechnology were beginning to open up ways of modifying existing viruses so that they would be more lethal, and could spread more quickly, than anything nature had made. Eventually, Block thought, it would be possible for biologists to design their own viruses—combining, say, the lethality of Ebola, which kills as many as 90 percent of its victims, with the communicability of influenza, which can spread with astonishing rapidity from one end of the world to the other. At the time, he said, the new biology was like physics during the Second World War: a kind of Pandora's box. Studies of the physical sciences had unlocked the power of the atom, creating a bomb that could unleash a thousand times more explosive power than conventional ones. A decade later, hydrogen fusion bombs multiplied this power another thousand times—a million times the destructive force of conventional weapons.[4]

The bomb that biology would make possible, though, was the infectious type: diseases caused by bacteria, viruses, and so forth. These new developments could lead to an increase in virulence proportional to the increase in explosive force of the atom and hydrogen bombs. Imagine a flu bug that's a thousand times more deadly than the flu of 1918, or that spreads a thousand times more rapidly. "The type of biological weaponry made possible by genetic engineering," wrote Block after the 1997 meeting, "may conceivably produce an analogous increase in virulence over conventional biological agents, which in turn have greater destructive potential than natural outbreaks of disease.

"If the analogy [to atom splitting] holds," wrote Block, "this is bad news indeed."[5]

Block let his mind wander even further into the dark potential of biotechnology. He thought of all the things that science was learning about the human body—all the biochemical "pathways" that control every fundamental life process, from cell division to cell death, from respiration to the transmission of a thought from one neuron to another. The National Institutes of Health were spending tens of millions of dollars each year on learning about these processes because, if they could get to the bottom of how diseases work, they might be able to devise cures. But as they say in the spy trade, this knowledge is "dual use"—it can also reveal ways of disrupting the process of life, turning it off like a switch. It could lead to bioweapons of startling effectiveness that work in ways we can only dimly imagine now.

Such threats could certainly cause terror, but could they threaten the survival of the human race? This isn't the kind of question that Block or anybody else for that matter can answer with any certainty, partly because we don't yet know where the science is headed, and what technological possibilities will soon emerge. But it's not hard to speculate what form such a threat might take. For a truly devastating weapon, you really need something that goes to the heart of our genetic vulnerability, something that has bedeviled the human race for as long as it began to thrive, something that is almost mathematical in its possibilities, something that you can aim right at the heart of the cell and the human genetic machinery. You need a virus.

Block wasn't saying that a powerful killer biological weapon was imminent. But he warned that bioterrorism was a growth industry.

> It seems likely that such weapons will eventually come to exist, simply because of the lamentable ease with which they may be constructed. In contrast to nuclear weapons, bioweapons do not

> require rare materials, such as enriched uranium or plutonium. They do not require rare finances: development and production are comparatively inexpensive. They do not require rare knowledge: most of the techniques involved are straightforward, well-documented and in the public domain. Today, thousands of biologists worldwide possess the requisite skills, and more are trained every day (most often at US universities). Finally, they do not require rare infrastructure; some bioweapons can be produced by small terrorist groups almost as easily as through national biological warfare programs.
>
> Inevitably, someone, somewhere, sometime seems bound to try something.[6]

Block thought it behooved us to start thinking about it. Genetic sequencing was about to go from an esoteric technique to a widely practiced method of research. It wasn't inconceivable that one day soon a terrorist would be able to download the genetic blueprint of a bad virus and concoct one in a lab without much more fuss than baking a batch of brownies. Perhaps it would be possible to manipulate such a virus to be even deadlier than anything we've ever experienced. Or perhaps even make something new from scratch.

Revisiting his ideas over a decade later, Block has only small revisions. With hindsight, a few details seemed not to have panned out, but on the whole, he was prescient. If anything, the techniques of biology have developed at a faster pace than it was anticipated in the 1990s. Craig Venter succeeded in creating an artificial life-form in 2010, inserting a prefabricated genome into a mycoplasma, a simple one-celled creature, and coaxing it to replicate.[7] It wasn't useful for anything except to demonstrate that it might be possible, one day soon, to build organisms the way we build factory robots, only these would be tiny and self-replicating; they could do much good, or ill.

Since Block's initial gloomy assessment, the field of synthetic biology has sprung up, with the goal of making biology into an engineering discipline. MIT sponsors an annual contest called iGEM, for International Genetically Engineered Machines, in which college students compete to build genetic "devices" out of "parts" they make in the lab—bits of organisms that are designed to fit together with other bits, like so many Tinkertoys, with knobs and sticks. Each time a team enters the contest the parts they make go into a library. The idea is to eventually be able to pick out parts from a catalog and use them to fashion all sorts of useful biodevices, quickly and cheaply.[8]

That's how it goes. Science turns into technology, and technology turns into products that we buy at Home Depot or RadioShack and use around the house and leave in the garage to get rusty.

All the incredible advances at the frontiers of biology and medicine seem destined to follow this path. This will give us the means to make life better, but also to do harm. "If we understand how to fight a disease, then we also understand what makes the disease so bad, and maybe somebody could figure out a way to make it worse," Block told me recently. "The only thing between us and some kind of biological Armageddon is the fact that for now at least these things are complicated enough that we don't understand them."

We'll see just how prescient Block turned out to be.

AS AN UNDERGRADUATE at Swarthmore, David Baltimore changed his major from biology to chemistry in part because he was less interested in birds and ecosystems than in what was going on inside the cell with big organic molecules such as DNA and proteins.[9]

As a young professor at MIT, Baltimore was intrigued by a class of viruses that did not use DNA as the medium for holding their genetic information. DNA is the genetic analog to Moses's tablets: the genetic information written in this molecule doesn't degrade easily. This molecule holds genetic information so that it can be passed from one generation to the next—by the standard of biology, it is stable. (DNA mutates from one generation to the next, but that's a different story.) Baltimore's curiosity drew him to viruses that were not made out of DNA at all, but were instead built of the much less stable RNA molecule.

How could a virus live and propagate and infect humans for hundreds and thousands of years without disappearing into the shifting sands of biochemistry, yet be built out of the molecular equivalent of gossamer? RNA's role in the cell's chemistry was thought at the time to be restricted to not much more than carrying messages from one end of the cell to the other and then collapsing like a runner at the end of a marathon. Yet many of the worst viruses—HIV, polio, influenza—are built of the stuff.

Baltimore zoomed in on the question of how an RNA virus, once it gets inside the cell, manufactures the DNA it needs to influence the cell's behavior. Viruses, like Mary Shelley's monster, are not quite alive, but they're not quite dead, either. They cannot reproduce without borrowing (and ultimately destroying) the cellular machinery of a host cell. Without a host, they cannot turn food into energy; they are a bag of inert chemicals. Baltimore discovered the enzyme that viruses use to convert their RNA into DNA inside the cell—an enzyme called reverse transcriptase. Baltimore shared a Nobel Prize for the discovery in 1975, at the age of thirty-seven.

Years later, Baltimore wondered if it might be possible to build his own viruses from scratch—from a scaffolding of genetic material assembled in the lab. Using what at the time was cutting-edge technology—a technology closely related to genetic sequencing, which allowed biologists to manufacture long strands of DNA molecules—Baltimore built a DNA-negative of the polio virus, kind of a molding cast for polio's RNA genome. If DNA's double helix is like a twisted ladder, RNA is a single side of the ladder, split in the middle, with broken steps of varying sizes hanging off it. Baltimore built a DNA molecule that had only a single strand—one leg of the ladder—and then used this single-stranded DNA to piece together RNA that matched it. In this way, he thought he might be able to piece together an entire RNA genome of a polio virus.

Baltimore strung together the DNA for polio and, using reverse transcriptase, built his RNA genome.[10] Then he injected it into mammalian cells. Once the virus entered a cell, it went into action, doing all the things polio viruses do when they infect a host cell to reproduce. A few hours later, the flask was teeming with polio virus—enough to pollute a drinking well. He published the research in 1981.[11]

Baltimore's motivation was benign. Being able to build polio viruses in the lab would help scientists greatly in coming up with better vaccines—they could build and test their own viruses, even different mutations, and see what worked and what didn't. It didn't take much of a leap, however, to imagine how a virus factory would come in handy for someone with less than noble aspirations. Baltimore's breakthrough had given bioweaponeers a blueprint of sorts,[12] but few people were thinking in those terms. The Soviet Union was the main worry—it had a big bioweapons program that had already developed a whole rogue's gallery of pathogens that could cause horrible diseases. But the United States and the Soviets signed the Bioweapons Convention by 1972, agreeing to limit their offensive bioweapons programs. The Soviets, we know now, continued theirs.[13]

In any case, Baltimore's technique wasn't exactly easy to pull off. It wasn't the kind of thing that some rogue nation could do. The operation required cloning, a delicate technique that at the time was brand-new, to produce the DNA mold. Only a few labs in the world had the equipment or the expertise to pull it off.

Twenty years later, the Soviet Union was no more, and hundreds of labs had equipment that Baltimore could only have dreamed of in 1981. Incredibly, the magnitude of Baltimore's feat—that he had actually synthesized a live virus from inert biochemicals—faded from the popular imagination. Until Eckhard Wimmer revived it in 2002.

Wimmer, a professor of biology at SUNY Stonybrook, brought people's attention to Baltimore's feat in a way that showed how far molecular biology had come in the past twenty years. He decided to show that it would be possible to build viruses the way you would make gunpowder or a fertilizer bomb—that is, inexpensively and even crudely. By then, a whole cottage industry of firms that manufactured strips of DNA and RNA had sprung up.

Wimmer and his colleagues sent away to biochemical mail-order labs for strips of DNA that they could use to assemble the polio virus in the lab, using a genome downloaded from the Internet. A few days later, the DNA strips arrived in the mail. Using techniques that had become routine for many biology labs around the world—including just about any college-level lab—the team came up with a beaker full of polio virus.[14] Then they held a press conference.

Wimmer was widely criticized for pulling off what amounted to a publicity stunt. The research showed nothing that was scientifically new.[15] But his point was valid nonetheless: in a world where biotechnological techniques were increasingly common, viruses are merely another form of information. The idea, said Wimmer, was that it was no longer possible to eradicate a virus permanently, because in principle you can always recreate it from the information that's out there. They are, in a sense, chemicals, like plastics or aspirin—something to be synthesized.

Bioweapons experts don't worry too much that polio is going to become a bioweapon. For one thing, there are vaccines for it. Except for some parts of Africa and Asia, the virus has pretty much been wiped out. Of course, that also means many people have lost their natural immunity to it. A polio virus released simultaneously throughout the United States might cause scattered outbreaks among people who hadn't been vaccinated. It might not hurt many people, but it could cause panic.

But worse diseases are out there—plague, for instance, or tularemia. Anthrax has already been used for the purposes of terror, of course, in the weeks after 9/11. The delivery mechanism was the U.S. Postal Service. At *Newsweek*, where I worked at the time, for months we opened all our mail in a separate room to keep confined any anthrax powder that might have been sent (none was). Anthrax is a nasty disease, surely, particularly when it's made into a weapon—that is, when it's turned into a fine powder that can leak out of the small spaces of an envelope, float in the air, and wind up in the lungs of innocent workers in postal sorting centers. But like many horrible potential bioweapons, you don't need a genome to get hold of it—it's out there, in the soil of dusty cattle ranches in the American southwest and elsewhere around the world.

What makes anthrax a good bioweapon, though, makes it a bad agent of Armageddon. The beauty of anthrax, from a military standpoint, is that you can target it to the population you want to reach, and it won't get out of control—won't spread further than the initial target of attack. The victims of anthrax poisoning, in other words, don't generally spread the disease to those around them. If you want to do a lot of indiscriminate damage, you want a particularly bad disease that can spread like wildfire, such as smallpox.

Smallpox is about as different as you can get from polio. Instead of 7,200 base pairs, smallpox virus has about 186,000,[16] and it's encased in a complex shell of protein that protects it from the elements and from the human immune system. Wimmer took a lot of flack for delivering his message so soon after the attacks of 9/11, which were only a few months before his press conference. He intended his feat as a warning about the potential of bioweapons. But folded into that message was a tone of reassurance: the smallpox virus was so complex, so different, that fabricating it in a lab seemed impossibly hard for the foreseeable future.[17]

But Wimmer's own stunt undermined this message. He showed that what was difficult for Baltimore had in twenty years become easy. The trouble with the foreseeable future is that it has a way of arriving sooner than you think.

AS THE PRICE of genome sequencing continues to drop, biologists have taken to sequencing the world around them, the better to understand how the enterprise of life works. By October 2009, the number of microbial genomes that scientists have sequenced passed the one thousand mark. They have publish the sequences of *E. coli*, one of many bacteria in our guts (and the occasional sprout or hamburger); the fruit fly, and the spotted green puffer fish of Asia. They've also sequenced many pathogens, the better to study them and figure out ways of fighting them.

Hundreds of human pathogens have been sequenced and their genomes made available to anyone who wants them. The Sanger Institute in England has funded many of these projects, and it operates a website that makes all the data public. You can find the genome for *Bordetella parapertussis*, or whooping cough, which was sequenced in 2003; it consists of 4,086,189 base pairs. You can also find progress reports on five different strains of *Staphylococcus aureus*, the staph bacteria that is resistant to many antibiotics. "This sequencing centre plans on publishing the completed and annotated sequences in a peer-reviewed journal as soon as possible," it says helpfully.[18]

Sequencing has become a standard tool for virologists looking for clues on how to fight human disease. When an Ebola virus swept through Uganda in 2000, killing more than two hundred people,[19] scientists at CDC went to work decoding its DNA, publishing their results in a 2005 paper in the journal *Virus Research*. They compared the virus to several other strains and discussed openly which genes differed in the various strains and what effect these genes might have had on the ability of the bug to infect humans and cause all sorts of disgusting symptoms. In the past ten years, scientists have sequenced Ebola, Marburg, and other filoviruses, which are some of the nastiest viruses ever to infect humans. Sequencing was used to identify the source of the recent post-earthquake cholera outbreak in Haiti in a matter of days.

Publishing a bunch of A's, G's, T's, and C's on the Internet isn't going to infect anybody, of course. What's scary is that it's now possible to use that information to build a nasty virus from scratch. Scientists already build pathogens in their labs all the time. Sometimes they delegate the task to their graduate students.

If making a polio virus, at 7,200 base pairs, from scratch is the biological equivalent of baking a box cake, making a smallpox virus, with 186,000 base pairs, is more like orchestrating an elaborate French meal for one hundred people, growing your own ingredients, raising your own cows, and stomping home-grown grapes to make your own wine. In other words, it's theoretically possible, but not likely. Still, every day it's a bit more likely, and the question comes back to this: when will making a killer smallpox virus be so easy that a disturbed individual can do it?

For twenty years, Ted Kaczynski, a Ph.D. in mathematics from the University of Michigan, holed himself up in a cabin in Montana and mailed homemade bombs to people who, in his fevered mind, were creating our technological society and thereby constituted a threat to freedom. He killed three people and injured twenty-three.[20] If another Kaczynski is out there, and he has a Ph.D. in microbiology, he probably can't manufacture a smallpox virus. At least not yet.

SMALLPOX MAY BE the most awe-inspiring human pathogen nature has ever invented. Compared to most viruses, it is a beast: variola, the virus that causes smallpox is encased in a fearsome shell of protein that makes it impervious to light, moisture, and other environmental hardships. A smallpox virus can ride a minuscule current of air for hours, drift through the streets, filter through the gaps in a windowpane, looking for a human host. It is in no rush to find one. The virus is so well protected against the elements that it effectively has no expiration date; a spore can lay dormant for years before someone comes along and breathes it into his lungs, where it settles in the soft tissue of the unsuspecting victim, enters the bloodstream, hijacks cells, and breeds. After two weeks the victim runs a high fever, and pustules form on the skin—little exit ramps for new virus spores yearning for another host. And the cycle begins again.

Smallpox was a fact of life for most of human history. It is a relatively recent virus, a relative of more ancient pox viruses of animals. When Europeans settled the Americas, it was variola, brought over by the first explorers like shock troops, that most likely wiped out Native Americans—the Mayans, the Incans, and other peoples throughout the two continents.[21] "By the end of the second millennium, it had killed, crippled, blinded or disfigured one-tenth of all humankind who had ever lived," wrote Steven Block.[22]

Smallpox is the only infectious disease that has successfully been eradicated. D. H. Henderson headed up the World Health Organization's smallpox eradication effort in the 1960s and 1970s, roaming the world in pursuit of outbreaks and vaccinating thousands of people. He was there for the last outbreak in Europe, in Yugoslavia in 1972, and continued to pursue it through Asia and the Middle East until the last known outbreak, in Somalia in 1977.[23] The WHO declared the disease officially eradicated in 1980. It was a masterful operation on a shoestring, and Henderson is credited with a talent for performing nearly impossible negotiations with local governments to allow his team in to administer vaccines and make all sorts of demands on local public health officials. He's a combination of Clara

Barton, General Schwarzkopf, and Gandhi, packaged in a tall, soft-spoken Midwesterner. In the biosecurity world, Henderson's reputation is second to none. When he speaks, in his quiet, folksy drawl, the room falls silent, like an old E. F. Hutton commercial.

Just because smallpox doesn't exist in the wild doesn't mean we can stop worrying about it. The United States and Russia kept samples, for reasons that make perfect sense: if the disease ever does pop up again, medical researchers would want live samples to test treatments and so forth. But the existence of these samples has caused some consternation. Do they really exist in only the two nations? Have other countries gotten hold of them?

It's one thing to have variola samples, quite another thing to have a bioweapon. How exactly would you get a vial of variola to the general population? But the Soviet Union had manufactured tons of smallpox that had been "weaponized"—reduced to a fine powder that could easily waft on minute currents of air, among crowds at an airport or stadium or subway.[24] The FBI and CIA have been paying Russian scientists to keep smallpox out of the hands of terrorists; there's no evidence that anybody has gotten hold of it, but we might not know until it's too late.

A smallpox epidemic would be indiscriminately destructive, not only to the target but also to the one who initiated the attack. It would be insane—suicidal—to initiate a smallpox attack. Why would anyone do such a thing?

In the pre-9/11 world, these questions were rhetorical. The attack on the Twin Towers changed that. Suddenly everyone began to wonder, what *would* happen if Al Qaeda got smallpox? This organization was bent on indiscriminate destruction. It was suddenly entirely plausible that a well-organized group of terrorists, with no affiliation to any state, unencumbered by the Geneva Conventions or even a survival instinct, would happily inflict smallpox on a world that no longer had natural immunity to the disease. The amount of vaccine for the disease was negligible.

Suddenly you could almost see how Al Qaeda would go about the operation. It would take small amounts of the virus to the four corners of the globe—perhaps concealed in a vial, perhaps not. Maybe the messengers would simply infect themselves and pass through nightclubs and subways and concert crowds as much as they could, spreading the disease by contact to innocent bystanders. Then hidden Al Qaeda leaders could sit back in their caves or compounds and watch the epidemic spread on the evening news. In a few weeks there would be news of an outbreak. Then there'd be news of several, and within a few days there would be general panic and quarantining, until it became obvious that quarantines were not going to be able to constrain the virus, that too many infections had sprung up in too many places. The next line of defense—vaccinating everyone within a certain radius, called ring vaccination—would not be an option, either, because there wouldn't be enough vaccine in reserve, and making more would take months.

Bioweapons experts had in fact simulated a smallpox attack on the United States only a few months before 9/11, called Dark Winter.[25] In this "tabletop" simulation, a group of experts played various roles in a situation played out according to certain assumptions. The attack would take place simultaneously in shopping malls in Oklahoma City, Philadelphia, and Atlanta with the kind of "weaponized" smallpox that the Soviet Union was known to have made. From thirty grams in total—an amount that would fit in the kind of pill bottle you might get for your arthritis medicine—three thousand virtual people took bits of the powder into their lungs. Each infected person in turn spread the virus to ten other people. The United States had a stockpile of about 12 million vaccines, the experts assumed, that were effective against this strain (though the Soviets are thought to have doctored smallpox so that vaccines wouldn't be effective against the "wild" version).[26]

Dark Winter simulated only one of many possible scenarios, but the exercise nevertheless provided an interesting look into how a smallpox disaster might play out.

The exercise took place over three days, but it covered the simulated events of the first two weeks after the attack. On day one, which corresponds to nine days after the actual attack, officials are informed of an outbreak in Oklahoma of twenty cases, with fourteen more suspected; the CDC sends one hundred thousand doses of vaccine. They realize that this is only the first wave—given the seven-day incubation period of smallpox, the next wave will come in a week or so. The question of the day is how to contain the disease. They decide to use the "ring" method, in which everyone who's come into contact with the initial victims is vaccinated, then all the people who came into contact with those people are vaccinated, with the idea of forming a wall of immunity around the disease, in the same way that firefighters cut down swaths of trees to halt a forest fire. Officials agree that the public should be informed, even at the risk of sowing panic.

By December 15 (day two of the simulation), the disease has killed three hundred people and infected two thousand in fifteen states, as well as

Canada, Mexico, and the UK. Canada and Mexico are demanding that the United States supply vaccines, but there's precious little to go around. The states vary widely in their ability to distribute the vaccines, and in some cases violence breaks out at clinics and against "minorities who appear to be of Arab descent." In the affected states, hospitals are overwhelmed with intakes, both from smallpox cases and others who are panicking over seasonal disorders such as colds and flu. Borders are closed to U.S. travelers. The outbreak dominates the news. False reports of cures circulate on the Internet. The federal government starts a crash program to make new vaccines, but it will take weeks—too little, too late. The president addresses the nation and asks for cooperation and implores people to remain calm.

By December 22, thirteen days into the epidemic, a thousand people have died, and cases have ballooned to sixteen thousand in twenty-five states—fourteen thousand having been added in the past twenty-four hours. The effort at ring vaccination clearly didn't work, and vaccine supplies are depleted (the Department of Defense is resisting calls for giving up its stockpile for civilian use). The spread is exponential. Ten other nations report smallpox cases. People all across the country, in all walks of life, are staying home, hunkering down, hoping they avoid infection. As a result, food shortages are growing. Residents in the infected cities are fleeing. People are demanding that the government confine smallpox victims and their immediate contacts, but how would you even identify them?

At the end of the two weeks covered in the exercise, the outlook appears grim. The CDC projects that by February 6—about the time a new vaccine supply is expected to be ready—as many as 3 million people could be infected, resulting in a million deaths. "These numbers are worst-case projections and can be substantially diminished by large-scale and successful vaccination programs and disease-containment procedures," says the report.

The Dark Winter report was criticized by some people as alarmist, particularly in the world-case projection of a million deaths. But it's not hard to imagine a scenario in which deaths are even higher. The exercise assumes that vaccine manufacturers churn out new vaccines without a hitch, which, as we've seen with the 2009 flu pandemic and at times with ordinary seasonal flu, is not a foregone conclusion.[27] And what about propagation of the virus abroad? What happens if the outbreak is quelled in the United States but returns like a boomerang from some other part of the world? That was beyond the purview of Dark Winter.

The Dark Winter exercise took place, spookily, in June 2001, three months before the 9/11 attacks. So it's no wonder that the Bush administration promptly authorized $428 million for 155 million doses, later increasing the stockpile to more than 300 million doses, enough to vaccinate everyone in the United States.[28] The stockpile is kept in secret warehouses throughout the country, near major population centers, waiting for word from the White House to send vaccines out by the truckload to hospitals and clinics.

But the vaccine is already obsolete. For one thing, it would be ineffective against Soviet-doctored smallpox. And the threat we're facing from here on out—a world in which more and more people can download genomes and tweak genes and build do-it-yourself organisms—is in many ways more difficult to contain.

There's no need to wait for this scary future. As far as smallpox is concerned, it has already arrived. It came several years before the nation, spooked by Al Qaeda, began amassing its smallpox vaccine stockpile the way a child arranges stuffed animals on his bed to keep the ghosts away.

IT STARTED AS a way of exterminating mice.

One of Australia's most vexing problems is pests. For the past few hundred years, the former colony has become home to a zoo of mammals—brought over to the island continent in the bowels of ships, or by well-meaning farmers—that have taken to their adopted land with an often grotesque fervor.[29] Feral cats, foxes, pigs, goats, toads, and rabbits crowd out or kill native mammals and reptiles and overrun or eat vegetation, causing soil erosion and turning the land into a desert. A fair portion of the energy and time and funding of biological researchers in Australia is aimed at finding ways of reducing these pests, which explains how Ron Jackson and Ian Ramshaw got interested in mousepox.[30]

Mousepox is harmless to humans, but to mice it is the face of Armageddon. The *Ectromelia* virus, which causes mousepox, replicates rapidly in the body of a mouse, and for the poor mouse that is born without resistance to the virus, infection means certain death. The virus quickly overwhelms the immune system, producing the lesions that characterize diseases in the pox family—similar to those produced by smallpox in humans. The lesions are little outward-bound populations of virus that collect in pustules on the skin, waiting to make the jump to another mouse upon contact. *Ectromelia* is a rather large virus as they go, with 176,000 nucleotides wrapped in a shell of protein that keeps it fortified against immune systems and the elements—sun, dryness, and so forth.[31] The death rate for mice that carry no resistance to the virus is close to 100 percent. Mousepox would make an effective pesticide against the rodents, except that most of the mice in Australia have acquired defenses against mousepox; many strains of mice are entirely impervious to the disease. They still catch the virus—a new strain of *Ectromelia* will still spread throughout the population like wildfire—but the mice who catch it just don't get sick. They carry on with their lives, breeding and thriving in the outback.

If Australians could wipe out their mice simply by sprinkling them with mousepox dust, they would probably do so happily. During the occasional "mouse plagues" in some parts of the country, mice can overrun homes, winding up in dresser drawers and cupboards and pillowcases. But Jackson and Ramshaw weren't interested in the lethality of mousepox, but rather in its ability to spread readily and quickly through the mice population. The two scientists had the idea to tweak the mousepox virus so that it could carry something else—a protein that would trick the mouse immune system into attacking the mouse itself.

But which protein? Jackson and Ramshaw took a protein from the protective gelatinous covering of the mice's own egg cells. (Human egg cells have a similar covering that admits a sperm, but keeps other things, such as potentially harmful bacteria, at bay.) A mother mouse's immune system knows not to attack her own egg cells in part because it recognizes this protein and knows to leave the egg alone. But what if you trained the immune system to react differently to this protein—to see it not as a friend but as a foe? If you could do that, you could persuade the mouse immune system to attack *its own eggs*, rendering the mouse infertile.

Typically a mouse has only a handful of eggs, and therefore the immune system only ever runs across a small amount of this friendly protein. The scientists figured, though, that if they could bring great quantities of this protein to the attention of a mouse's immune system, it might have a change of heart and come to see the protein as an invader. It's sort of like eating too much fudge. A little bit is delicious, but if you indulge over and over, to the point of getting sick, the sight of those little brown squares is enough to make you nauseous.

So that was the plan. Take a mousepox virus, attach a protein from the gelatinous covering of mice eggs, infect the mice, let the virus replicate (and the protein as well). The grossed-out mouse immune system would take care of the rest.

Ramshaw and Jackson developed just such a virus, injected it into their experimental mice, and waited. Nothing much happened.

Some of the mice died, which wasn't what they were hoping for. The idea wasn't to kill the mice, but render them sterile.

Many of the mice didn't die, however. After doing autopsies of them, the scientists found that the survivors had such strong defenses against mousepox that their immune systems neutralized the virus before it had a chance to replicate in the bloodstream. Thus the immune system never saw much of the protein.

So the scientists went back to the drawing board. They added another ingredient to their engineered mousepox virus that would suppress the immune response, so the virus could proliferate enough to let the immune system recognize the protein and turn against it. The ingredient they added was interleukin-4, a chemical of the immune system that blocks the production of antibodies.

Once again, they injected mice with the revised virus and waited to see what would happen. They hoped and expected that the virus would kill a few mice, but leave most of them alive—and that these survivors would be rendered sterile. But, alas, that's not how it worked out. When they next looked at the mice, every single one of them had died.

This was puzzling.

The mice the scientists were using were known to be resistant to mousepox. The IL-4 shouldn't have suppressed the mouse immune systems enough to leave every last mouse vulnerable to the virus, but perhaps something else was going on. The researchers needed more information. They decided to run the experiment again, but this time, they would vaccinate the mice against mousepox before infecting them. If the pox virus had indeed killed the mice, this time it should leave them be. The vaccinated mice should be able to handle the new virus without too much difficulty.

When they checked the cages, they again found dead mice—every last one had keeled over.

The scientists then realized what they had done. In their efforts to build a better mousetrap, Ramshaw and Jackson had stumbled on an idea for a potentially devastating bioweapon. By adding IL-4 to the mousepox virus, they had created a new virus that could kill previously vaccinated mice—with a mortality of 100 percent.

Mousepox is harmless to humans, but it bears a strong similarity to smallpox, which is not harmless to us at all. Most of the world's nearly 7 billion people have no resistance at all to smallpox; contracting the disease would be fatal to a large proportion of them. There are vaccines against smallpox. But would it be possible to add IL-4 to a smallpox virus and come up with a biological weapon that would be impervious to known vaccines? With a kill rate of 100 percent?

Jackson and Ramshaw published their findings in the *Journal of Virology* in July 2000 without mentioning the link to bioweapons. That came up when a reporter, Rachel Nowak, from *New Scientist*, interviewed Ramshaw about his work with AIDS, and he casually mentioned it, almost as an afterthought.

WE KNOW FROM Jackson and Ramshaw and the work of other scientists who have had similar results that it could soon be a fairly simple matter to take ordinary smallpox and tweak it so that it turns extremely deadly. And we know that disseminating a weaponized version of smallpox would not be difficult—all you'd need to do is let the spores out near the air vents of half a dozen major airports at once. But to modify smallpox and release it to the world at large, you first need to get a sample of the variola virus. So far as we know, those are locked away and under guard, safe and sound, in the Vector Institute and the CDC (and perhaps in a few other places). Let's assume, for the sake of argument, that to get a smallpox virus, you'd have to make one from scratch. The smallpox genome was published and is now in the public domain, so that's not a problem. However, to make one in the lab, you would have to overcome two problems.

First, you'd need to be able to make a large genome. It's not easy, but it's getting easier each year. Craig Venter's genome for mycoplasma, a simple bacterium, has a whopping 580,000 base pairs,[32] compared to variola's 186,000. Venter may have one of the better-equipped labs around, but we know that today's cutting edge is tomorrow's high school norm.

The CDC administers the Select Agents and Toxins List, which includes, among other things, key genetic sequences that are characteristic of pathogens such as smallpox, Ebola, Marburg, and other viruses that could attract the interest of someone wanting to make a bioweapon. The select agent program is also called, colloquially, Black Watch. Like any safeguard, it's not foolproof. A Luddite Unabomber with a Ph.D. in microbiology with the nerve to order smallpox in five easy pieces might not get away with it. But a clever Unabomber might know enough to slice up the variola genetic sequences in a way that doesn't set off alarm bells. The companies who supply the sequences—Blue Heron Technology, Codon Devices, DNA

2.0—aren't required by law to enforce oversight of the CDC's select agents list. They are motivated by a fear of liability should one of their customers use their products to build a horrific biological agent that winds up killing lots of people. "They're doing it out of enlightened self-interest," says Block. "Who knows what criteria they follow. Some of them say they do check for these things, but maybe all of them don't."

Using mail-order sequences wouldn't break the bank, either. Prices are currently running less than fifty cents a base pair, says Block. A few years ago, the maximum length you could order was about forty thousand base pairs, but that's risen to one hundred thousand, and will no doubt rise further in the next few years. "The order-a-virus-by-mail era is upon us already," says Block.

A nation that wants to make smallpox could simply use its own DNA synthesis outfit and set whatever rules it wants. These days most major nations have their own bioweapons labs, and in any case they have the equipment to experiment with smallpox. I have spoken with many bioweapons experts with knowledge of which nations possess labs that adhere to biosafety levels three and four—those with air locks and reverse pressure systems and so forth designed to contain nasty bugs so it's safe to experiment with them—and they have been reticent. But a few referred me to Wikipedia, whose list includes Australia, Belarus, Canada, the Czech Republic, Finland, France, Gabon, India, Indonesia, Italy, Malaysia, Poland, Singapore, and Spain. Other nations may not adhere to the safety protocols but still have labs equipped to play with the nasty bugs. There are rumors that Syria has such a lab and smallpox samples too, but this is not a mainstream view.[33]

But there's still one problem. Viruses that are made up of RNA, such as polio, have self-contained genomes, which means you can just string together all the base pairs, implant the molecules into a cell, and bingo—the virus comes to life and does what a virus does, which is to say it takes over the cell and replicates. But a challenge for the would-be terrorist making a smallpox virus is that even if you assemble all the DNA into a full-fledged smallpox genome, it wouldn't be easy to bring to life. When Wimmer made his polio virus, all he had to do was piece together the RNA genome, insert it into an empty cell, and wait for the virus to breed. But smallpox is not made of RNA, it is made of DNA. Assembling the DNA base pairs into a variola genome and implanting it into a cell would amount to nothing but a dead hunk of DNA. It would not come to life, it would not take over the cell, it would not replicate, and you would not wind up with a culture teeming with smallpox virus. You would need to add enzymes and a bunch of other ingredients to the culture to give it the spark of life. These extra ingredients are packaged up in the variola virus, along with its DNA, inside the protein wrapper. Getting that concoction right in a lab would be a difficult job indeed.

On the other hand . . . when it comes to biology, there's always another hand—an alternate route. It's possible to take some other genome and pull it and push it until you get it to change into the genome you seek—to take genomes of mousepox or bird flu and turn them into pathogens that humans should worry about. The technique for doing that is called reverse genetics.

Reverse genetics illustrates perfectly the "dual use" nature of microbiology: anything that does good can also do harm. It was invented a few years ago in an effort to develop flu strains for testing vaccines. But it could also provide would-be bioweaponeers with another tool in their kit.

PETER PALESE, ONE of the world's most innovative flu researchers, developed reverse genetics in the 1990s for reverse engineering an influenza virus.[34] Essentially, it allows scientists to replace a passage of RNA in an influenza virus with a stretch of RNA they manufacture in the lab. It sounds like gene splicing, but there are some key differences. For one, reverse genetics can be used with RNA, which scientists previously had no way to manipulate, as they do so easily with DNA. But it also allowed scientists to change many passages of genetic code at one time. They can effectively take a starter genome and add and replace many sequences at once, sort of like taking a trailer home and adding so many additions that it's transformed into a palace.

Gabi Neumann, a microbiologist in Kawaoka's Wisconsin lab, took Palese's technique and found a way to use it to modify as much of the influenza genome as needed. Her method is to attach a passage of RNA that she wants to insert into the virus to a ring of DNA called a plasmid. Then she injects the plasmids into a cell, and the rest takes care of itself: the virus takes over the cell and replicates as normal, but adopts RNA from the plasmids.

The technique allows scientists to essentially design any influenza virus they want—even those that have never appeared in nature. Reverse genetics is being developed as an alternative to the chicken egg in the manufacturing of a vaccine, but it's not quite there yet.[35] Rather than letting viruses replicate willy-nilly, then sifting for the strain that would make a good vaccine, reverse genetics allows scientists to directly manipulate the virus's genetic makeup. (Reverse genetics, developed for influenza viruses, could be used to turn a mild flu bug into an effective bioweapon.)

That pretty much takes care of the two big obstacles to a smallpox bioweapon. Says Steven Block, "The two technologies that are required to resurrect smallpox from the grave so to speak are the reverse genetics system to make the virus itself from the DNA, and that's possible in principle now thanks to recent developments. And the other would be the ability to synthesize a DNA segment that long, and that's now also possible thanks to recent developments."[36]

Mail-order companies that supply DNA sequences use a computer program called Blackwatch to identify sequences that could conceivably be assembled into the CDC's select agents. Blackwatch, says Block, is "the only thing that stands between us and someone being able to make smallpox."

SMALLPOX IS ONLY one pathogen. Block has a laundry list of potential threats that he sees coming down the road, as biologists find new techniques and those techniques trickle down from the elite labs to the commonplace ones.

Gene therapy is a way of modifying a person's genome by sending new genetic material into his or her cells. It's an experimental technique for repairing or replacing a faulty gene that is causing some illness—diabetes, heart disease, cystic fibrosis, muscular dystrophy, cancer, mental illness. The idea is to piggyback the gene you want onto a deactivated virus, which takes it into the bloodstream and deposits it inside target cells like a Trojan horse carrying troops. Pox, herpes, and cold viruses are often preferred. Some scientists believe it may be possible to deliver gene therapy with a nasal spray, which for nefarious purposes could be an atomizer in the subway.

You don't have to look far to get an idea of how destructive gene therapy could be as a bioweapon. It's dangerous enough as therapy. In 1999, eighteen-year-old Jesse Gelsinger took an injection of a deactivated cold virus into his hepatic artery, which leads to his liver. The virus carried a gene that was supposed to replace a defective one,[37] but the patient died. Three other patients acquired leukemia from similar treatments. Scientists haven't given up hope for the technique as a form of therapy. Unfortunately, neither should weaponeers.

Stealth viruses. It may one day soon be possible to design a virus that operated in secret, infecting populations without causing any apparent symptoms. Humans already carry quite a few viruses in our bodies that don't have any effect on us—they don't cause fevers or give us headaches or runny noses, but simply exist, and we're unaware of them. Many people now carry herpes simplex viruses but don't develop cold sores, or the Epstein-Barr virus without developing mononucleosis. Sometimes these viruses do nothing until they are activated by some environmental stimulus—oral herpes, for instance, hangs out in facial nerves until something happens to awaken it. A virus can be triggered by sunburn, a cut or scrape, or by another unrelated virus, or physical or mental stress.

You can imagine an evil microbiologist designing a stealth virus that spreads throughout a population, lies dormant in people's cells, and springs to action only when the microbiologist sends out some trigger—perhaps a second virus that activates the first. You can imagine what would have happened if Osama bin Laden had infected the entire world with a stealth virus, then started to make demands on threat of triggering mass deaths. If that sounds too much like science fiction, consider that the biotech firm Invitrogen currently markets a product called GeneSwitch that can control the expression of genes in lab rats. Let the rats grow up, then—bam!—trigger the onset of heart disease or cancer. To be sure, it's a leap to call GeneSwitch a tool for world domination, but it shows that the concept of triggering gene expression is hardly far-fetched.

Escape viruses. Many of the worst modern human pathogens emerged from primates. Ebola is thought to have jumped to humans from bats, HIV may have come from chimps, influenza viruses arise from birds. But many more viruses are all around us—in the dog in the backyard, in the cat on the windowsill, and in each and every bird and mouse and opossum. The tropics hold more viruses than scientists can count. They only bother us when they find themselves outside their normal realm—when people move into rain forests and start cutting down trees and come into contact with the animals who host these exotic viruses. Most of the time an animal virus either has no effect on humans or it causes no serious illness. But once in a while, with the right circumstances and an opportunistic mutation or two, you get a scary new pathogen. Some of these diseases require no fiddling. Intelligence experts fret about the biological equivalent of a suicide bomber wandering the New York City subways spreading Ebola.

This situation suggests a strategy for the right kind of bad guy. By introducing a genetic change or two to a benign animal virus, it is theoretically possible to create a new human pathogen, perhaps one that is lethal. Canine parvovirus illustrates the idea. Parvovirus was not known to be a disease of dogs prior to 1974. At that time, however, a virus that causes distemper in cats—feline panleukopenia virus—acquired two minor mutations.[38] Nobody suspects foul play—viruses are mutating all the time, and every once in a while a mutation will make a virus suitable for another host. The mutated cat distemper virus turned into a disease that kills nine of every ten dogs infected. A vaccine is available for it, fortunately—and it would be a good idea to make sure your dog has been inoculated.

Canine parvovirus was certainly a "naturally emergent disease," as the CDC puts it, not a willful act of malice. But what if the next one isn't? And what if instead of infecting dogs, it infected humans, killing nine of ten—a lethality rate similar to the worst strains of Ebola? It might take months to identify it and develop a vaccine, and for many people it would come too late.

Designer diseases. The scariest prospect of all is that some expert terrorist could devise a killer virus from scratch—the prospect that someone could first contemplate the desired effect, and then fashion a bug that produces that effect in the human population.

Another possibility is to combine the worst qualities of existing diseases into one supervirus. For instance, influenza is highly transmissible. Flu bugs often spread from one end of the world to the other in weeks. But they often aren't particularly virulent, and don't cause many deaths. Ebola, on the other hand, is highly virulent, killing 90 percent of victims in some strains. Generally there's a trade-off between these two qualities: a highly lethal virus tends to kill its victims so quickly that they die before they can pass the virus along to the next person. But perhaps there's a way to make a lethal disease that doesn't kill right away—that disguises itself as a common cold for several weeks, while it gets passed along on the subway and in the mall and wherever people come into contact with one another, and only later turns deadly.

The missing ingredient here is the knowledge of how diseases work at a genetic level. What are the genes that make a disease easy to transmit?

Which genes make it deadly? Scientists are only beginning to study these questions, but the more they discover for purposes of cures, the more the bad guys know for purposes of biowarfare.

So what happens when this kind of knowledge becomes available? The prospects quickly get scary. A designer bug might focus on the immune system, much like HIV. Or it might stimulate dormant genes we already have to wreak havoc on our bodies. Every cell in our bodies contains a suicide mechanism that shuts down all functions and leads to a quick death. It's called apoptosis.

As you'll recall from high school biology, apoptosis is perfectly natural, even necessary, for normal development. A fetus, for instance, goes through many forms in the womb. At one point it has webbed feet, but later the webbing simply goes away and the toes form. The cells that make up the webbing basically kill themselves, using a genetic program that each cell has for committing suicide. Programmed cell death is used to pattern our brains, pruning neural pathways that have proliferated when we're infants.

Apoptosis is one of those fascinating physiological mechanisms that has long been known—it was discovered 150 years ago—but it went largely unappreciated until the 1980s. Scientists studying follicular lymphoma noticed that patients suffering from this form of cancer had a mutation in their chromosomes—a bit of chromosome 14 was stuck onto chromosome 18, which causes a gene, BCL-2, to turn on, when normally it remains off. BCL-2 inhibits cell death. So when BCL-2 gets turned on, a cell will be less likely to initiate apoptosis. Cells with inhibited apoptosis are more likely to grow uncontrollably, forming a cancerous tumor.

Since then, scientists have identified many other genes that play a role in apoptosis. Worms, for instance, have a gene called CED-3, which turns on when cell death begins. CED-3 has an analog in the human genome—a gene that produces protease, an enzyme that slices DNA molecules into bits. Think of it as a kind of demolition gene that turns a living cell into a sack of genetic garbage. Scientists have found apoptosis genes common to all life-forms, from hydras to insects. The mechanism evidently evolved more than a billion years ago, when life was still young on the planet.

Drugs that make use of this relatively new knowledge are undergoing testing. Abbott has been working on a drug that inhibits BCL-2 as a treatment for follicular lymphoma.[39] The idea is to target a gene that is a root cause of cancer, rather than carpet bomb every cell that grows with chemotherapy. Inhibiting BCL-2 would once again give the cell the ability to kill itself using its natural apoptotic programming, which might make a follow-up course of chemo all the more effective.

"Knowing that BCL-2 plays a role in regulating cell death does not unlock this process to biologists, because it's grossly incomplete," says David Vaux, a biochemistry professor at La Trobe University in Victoria, Australia. "It's a bit like trying to figure out how a car engine works, and finding that spark plugs played some role. The trouble is, you still don't know that it runs on gasoline, or that it has pistons and uses coolant and a fan belt and so forth."

Keeping cells from killing themselves is a major part of the tool kit of many infectious agents. When a virus invades a cell, one of the first things it does is neutralize the cell's apoptosis mechanism. The idea is to keep the cell alive long enough so the virus can take over and hijack the cell's machinery to reproduce. Once this is done, sometimes a virus will initiate cell death as a way of escaping the confines of the cell, to go on its way in pursuit of other cells to infect. Adenoviruses, which cause colds, pneumonia, and other respiratory illnesses, use cell death inhibitors in this way, and so does Epstein-Barr virus, which causes mononucleosis.

The process of cell death is particularly scary because it is so fundamental to life itself—it's an evolutionary tactic that cells acquired at the very beginning of life on the planet. As scientists learn more about this process—as they begin to uncover all the genes and proteins that play a role—they are also unwittingly identifying ways for bioweaponeers to attack. What if the mechanism of apoptosis were harnessed as a biological weapon, if a drug could push the cell's suicide button, activating the apoptosis pathways? What if you designed a virus, say, that could infect cells and apoptosis, telling all the cells in an organism to kill themselves simultaneously, promptly and efficiently, with the force of a billion years of evolution behind them?

A virus that could get into the bloodstream, replicate widely, and then, after it pervaded every organ, activate the apoptotic pathways that already exist—that nature designed into each cell—would be devastating beyond anything we've seen in the worst earthly pathogen.

I asked Steven Block if we know how to make an apoptosis bioweapon.

"Oh, absolutely. We know a great deal about how to do that. We know most of the major genes that are involved in the apoptosis pathways. I have a catalog cover I got from a company called KevaCon International, from 1997. They were selling products to biologists to study. The title of the catalog is literally 'Death by Design.' And what they marketed were products that activated the apoptosis pathways, and they marketed these things to the cancer biologists studying this stuff. So we now understand the proteins and the signals that are used."

You'd have to be able to take what we know about apoptosis and translate it into a virus or bacterium that would be able to infect people, to go from one person to the next, and then at some point activate these pathways. But not too soon, or the agent would be too virulent to spread widely. Says Block, "It's a long way, thank God, from there to actually programming some [virus] or some bacterium to synthesize whatever is required to activate these pathways and get it to travel from person to person—to make a disease out of it. There's a whole long set of things that have to happen.

"Many of which, by the way, I could outline, but I'm not going to do it." Block followed this statement with a grim laugh.

How long Block can keep this secret is anybody's guess.

EACH MORNING, SHERAN Hussein's mother drove him the twenty miles through winding country roads to Gaston Day School, where he was a junior, and then back again in the afternoon. One morning in November 2008, Hussein, a junior, was scanning the radio for a catchy song, or an interesting tidbit—anything to alleviate the tedium. Then something on the local NPR station made him prick up his ears.

"Have you heard of the bacterio-clock? . . . How about bactricity? Maybe a bacuum-cleaner? . . . Well, they have one thing in common: they are all synthetic biology projects that whiz-kid undergraduates have dreamt up for the 2008 International Genetically Engineered Machine competition, the iGEM."[40]

Hussein stopped channel surfing.

The reporter went on, "The idea is to use a toolbox of biological parts—like different snippets of DNA, for example—to build a machine that operates inside living cells, inside bacteria, inside yeast. And using this approach, students are making cellular biofuel factories, they're devising organic sensors, even solving simple calculations with bacterial computers."

Hussein knew that DNA was a long molecule—a double helix, like a twisted ladder—made up of a series of four basic nucleotides, which acted as the "letters" of the human genetic blueprint. He knew that a gene was a stretch of DNA that acted as a unit, often conferring a specific trait, and that the genes were gathered up into the twenty-three chromosomes that make up the complete human genome. He was aware that scientists had already sequenced the nucleotide letters of a human genome a few years before, when Hussein was in elementary school, and that they celebrated the victory in a ceremony on the White House lawn. He might even have remembered from biology class what a plasmid was. (It's a ring of DNA that scientists sometimes use to ferry molecules into a cell.) But he had no experience in a lab, outside the superficial exercises required for his AP biology class. He didn't know what all these genetic doodads had to do with turning a living cell into a machine that could be bent to some purpose (for good or ill). He wasn't sure what synthetic biology was, but he liked the sound of it. Making biological machines that operate inside living cells—how cool was that?

What a great idea for a senior thesis, he thought.

Before taking this to his teachers, Hussein decided to put in a phone call to one of the organizers of the iGEM event at MIT. He started asking questions and quickly realized it was better to confess right then and there to being a mere high school student. If this wasn't going to work out, he might as well know now. This is where Hussein's people skills came in. On the phone, he sounds older than his years—articulate, enthusiastic, without apparent self-doubt. To his surprise, the organizers didn't discourage him from entering the competition.

The next thing Hussein did was bone up. He needed to arm himself with some facts. As a start, he did what any aspiring synthetic biologist would do: he googled *synthetic biology*.

By late 2008, coming up to speed on the latest techniques for manipulating the genetic stuff of life did not require a Ph.D. or a security clearance from the CIA. Hussein quickly found a primer, downloaded it to his thumb drive, took it to Staples, and asked them to print it out. It weighed in at seventy-five pages. "Some grad student had done it," he says. "Basically it explained what synthetic biology was in a very easy-to-understand way. It didn't tell you about the lab processes, but it went into the basics of how plasmids work, and how to insert foreign genes into plasmids. The concept." Hussein spent evenings and weekends poring over the document. "I highlighted everything."

For a subject as complicated and technical as molecular biology and genetics, the concept was astonishingly, alluringly simple. You take a single-celled organism—some yeast, say, or *E. coli*, the common gut bacterium—and tweak its DNA in such a way that the cell will do something you want it to do. Perhaps you want it to turn pink when the temperature exceeds fifty degrees Fahrenheit. Or you want it to glow in the presence of some toxin.

You aren't exactly making a new organism from scratch, but you're coming awfully close. You're taking a living cell, which evolved according to natural selection, and tinkering with its genetic machinery, adding parts here, snipping out parts there, using a tool kit that the contest organizers would provide.

Hussein took the idea to Ann Byford, his biology teacher at Gaston. She had worked in a pediatric cardiology lab at the University of Virginia as a lab technician, much of that time charged with maintaining a colony of transgenic mice, so she knew her way around a lab.

Byford handed Hussein a textbook on *E. coli*. "Get to know your friend," she said.

The practitioners of synthetic biology—the science of creating living things to perform chosen functions—are not mad scientists bent on destroying the world. They are just smart people who are trying to solve problems—more engineer than Dr. Frankenstein. Synthetic biology is to Hussein's generation what the crystal radio was to those growing up in the middle of the twentieth century. The first step in this direction is to make a biological version of LEGO—so that building a synthetic organism is a matter of fitting standardized parts together, so that almost anyone can do it. These biological LEGO pieces are called biobricks. The iGEM competition was designed, in part, to promote the idea of amassing a registry of biobricks, and in part to get groups to add to the biobrick catalog with their own parts.

When Hussein and a handful of his classmates entered the iGEM competition, they got a package in the mail from MIT with biological parts—a hundred or so biobricks—to use as raw materials for their invention. The biobricks were tiny plastic bags full of dried powder—bits of DNA—and some live samples of *E. coli*.

"We put up the smart board, and the kids got out their laptops. Everybody was looking at things, throwing ideas out, and we figured out everything that was possible to do," says Byford. One of the students had an idea for a deodorant with bacteria that could absorb the worst underarm odors. "I had to tell them, do you really think people are going to spread bacteria under their arms?"

If they chose something that would be seen as useful, they figured, it might be easier to raise money. They got the germ of their idea from one of the past entrants, who had found a "nitrate sensitive promoter"—a section of DNA that would kick into action in the presence of nitrate. This was a good idea for several reasons. For one, it had a big local value. Nitrates from the livestock farms of North Carolina were a big polluter of groundwater. The idea was to make a bacterium that you could drop into a glass of water, and if the nitrate level was above a safe amount, the bacteria would glow. The ability to detect high nitrate levels quickly and easily in drinking water could be useful.

One of the things that wasn't included in the kit from MIT was a plasmid that would be able to deliver this specific gene to the *E. coli* bacteria that the Gaston group was going to use. Byford had to look for someone that had done that work already, and which they could piggyback on. She did some research in the scientific literature and found a professor at the University of California in Berkeley who had already isolated a plasmid for the *E. coli* they would be using. She sent him an e-mail. A few days later, she got an envelope in the mail from Berkeley with some powder—dried plasmid—in a plastic bag.

To get this particular plasmid to work, Byford realized the kids would have to slice it up, edit out some of the sequences, and put it back together. It could be done by using a molecule that was sensitive to nitrate—one that reacted chemically to nitrate molecules—and could be used as a kind of chemical detector. The trick would be to attach another molecule to it that would cause the compound to glow in the presence of nitrate, and then attach the whole kit and caboodle to the DNA of an E. coli bacterium. The result, if all went well, would be a test for high nitrate levels. Drop a bit of *E. coli* into a cup of water, and if there was too much nitrate, it would glow green.

At first the task seemed daunting. Byford reckoned that the students would have to scrape together about $15,000 in materials and equipment, in addition to the kit that MIT had supplied. That seemed like an inordinate sum. Hussein and his buddies went around to local businesses to see if they would sponsor the project, but the reception was cool. Some of the prospective sponsors gave the boy quizzical looks. "I learned to anticipate their fears," he says, "as though I knew they were thinking, 'Your name is Hussein, and you want to put bacteria into drinking water?' "

With the summer approaching, it quickly became apparent that the kids were going to have to find innovative ways of cutting corners. One way was to make do by making their own equipment. They had the means of cutting and splicing DNA—that required some enzymes, some of which the school lab already had, and others that came in the iGEM kit.

Cutting up DNA with enzymes, though, basically gave you a beaker full of a cloudy liquid—that's what DNA in water looks like. That alone didn't do you much good.

What they needed was a way of sifting through this slurry and separating out the molecules they wanted from the rest. This required them to "visualize the DNA." The idea was to pour the DNA slurry onto a gel, through which they would run a small electric current. The current would push along the DNA molecules, but only the smallest ones would make it to the other end of the gel. The middle-sized molecules would get stuck in the middle, and the biggest ones would hardly get anywhere at all. What they would have, at the end, was a block with rainbow shades of DNA, each color band corresponding to a different-size strip of DNA. To get the rainbow, you needed to mix the DNA with a special dye that would show its colors under ultraviolet light.

This setup is called a white box, and they typically sell for thousands of dollars. Hussein and his team didn't have that kind of cash, so they improvised. They got a piece of glass, assembled a box, and mixed their own gel.

Even with the setup complete, the group was having trouble. They would take the plasmids they got from Berkeley, mix them into a solution, add the enzymes, and run the slurry through the gel, which took hours. Then they'd flick on the UV light with great anticipation, only to see nothing but a monochrome block of gel. The July 4 weekend came and went, and still no color bands of dye emerged. The group began to get discouraged.

One morning they met and seriously discussed whether they should proceed. The work would have to continue through the academic year if they were going to be ready by the November deadline. Did they really want to do it?

Then one day, after they had run so many trials that nobody could remember how many, one of the kids flicked on the UV light switch, and bands of dye appeared. "We have bands! We have bands!" the kids started yelling down the corridors.

"I was thinking that if any of the staff had heard us, and they didn't know what we were doing, they would have thought we were crazy," recalls Hussein.

The group had crossed a crucial milestone to fashion their biobricks. They were now able to separate out the stretches of DNA in the plasmid from Berkeley. What was left now was to attach the standard biobrick sequences at the tips and reassemble the plasmid.

In the end, the Gaston group ran out of time. They never managed to insert their gene into the *E. coli*. They went to MIT in November anyway and made a presentation. Of the 130 teams that competed, 30 had not pushed their projects through to completion—so they weren't in bad company. When they presented their work to a panel, one of the professors who directed the project started asking all sorts of questions, Hussein recalled.

The group had earned a victory of sorts. They showed that it was possible for a handful of bright high school students, with a little guidance from a person with lab-technician experience, to do what was nearly impossible for anyone to do a few decades ago. The total budget for the group came in at less than $2,000.

Ten or twenty years from now, what will high school students be able to do? How many items on Block's fantasy list of bioweapons will they be capable of concocting?

CHAPTER 6

Machines

Tara Estlin was a toddler in the 1970s, when artificial-intelligence gurus were talking about how computers would one day soon act as personal assistants, dashing off letters or booking travel arrangements or doing a bit of research. In retrospect, the idea seems naive, hatched in the minds of men (they were almost all men) in the days when office buildings held row after row of desks, where women sat doing work now done by PCs and Google.

The idea of a coherent field of research whose goal was to produce ubiquitous digital secretaries began to decline at about the same time as did ubiquitous human secretaries. By the time Estlin became a Ph.D. candidate in AI twenty years later, the subject had splintered into half a dozen different fields—or, perhaps more accurately, programming techniques. It may be a rule of technology that the more dull sounding an idea is, the closer it comes to actually changing the world. Artificial intelligence never really got there, but "partial-order planning" is already making waves. That is the field Estlin took an interest in, even though it sounded like a geek parody.

Don't laugh too hard, however. Partial-order planning could take us down if the geeks aren't careful.

Estlin was well aware of how misleading a name can be. For her Ph.D. thesis at the Unviersity of Texas, Austin, Estlin came up with a better method of writing computer programs whose purpose was to plan. As she wrote:

"Planning is a ubiquitous and integral part of everyday life. We plan when we go to the grocery store, when we take a trip, and even when we take the dog for a walk. Small tasks take very simple plans, such as driving to a restaurant for dinner, while larger tasks can take very extensive and complicated plans. For instance, building a new house requires a set of long and detailed plans to be done correctly. Planning also has an important role in the workplace . . . for . . . building pieces of machinery or designing efficient delivery routes. Much time and effort is often required to construct correct and efficient plans for such problems and many human hours can be spent on this process."[1]

This is not to say that Estlin wanted to write better computer programs for walking the dog. But tremendous power lies in getting computers to handle more ordinary tasks, and indeed the computer industry has already taken us down this path: computers have begun to exert control in virtually every sphere of our lives. The machines of industry have computers embedded in them. Our coffeemakers have them. Our gas and electric meters have them. Our cars have them. And the next step is to give them more sophisticated tasks. Many cars, for instance, can take over rapidly in the moments preceding a collision, the better to avoid it. To do this, the car has to make decisions. When does it take control? And when it does, how much should it apply the brakes to each wheel, and in what order? And when should it allow the reflex-challenged human to regain control?

The computer in your car has to make these decisions on its own. It wouldn't do to have it ask you, in the split second before impact, whether you want to deploy the air bags. The utility lies in autonomous decision-making.

Autonomous is exactly what computer scientists call machines that can make their own decisions, without human intervention. Estlin and other computer experts are trying to extend the ability of computers to act, not just to decide when a car is headed into a spin so it can put on the brakes, but also, say, how to drive from your house to the local Big Bear without getting stuck in traffic or getting into an accident. A big reason that cars on the streets cannot chauffeur us around town is that they can't make the myriad decisions they would have to make along the way. Should I go in reverse or forward? Is someone in my way? Should I brake for that little squirrel in the road even though someone's tailgating me? And so forth. The challenge for people in Estlin's profession is to figure out exactly how to prepare computers to take on such tasks.

One of the big drawbacks to computers is that you have to tell them exactly what to do. If you want a computer to drive you to the nail salon, you couldn't write a program explicitly for that task. Or, well, you could, but it would be more expensive and certainly more time-consuming than hiring your own driver, and if something changes, the program would be too brittle to adapt. A computer program is like a dutiful but dull employee whose boss has to keep telling him exactly what to do, how to do it, and when, each time he finds himself in a slightly new situation. A good boss will give his employee a goal, some guidance, and the tools he needs. If a problem comes up, you want your employee to be resourceful enough to get the job done, but have the judgment to know when a problem requires action that exceeds his authority or expertise. Programming an autonomous computer is an exercise in optimizing all these kinds of decisions. Turning computers into good, resourceful employees is precisely what Estlin was trying to do.

The trick is to design a program that enables the computer to move toward a goal even in unpredictable circumstances. Estlin set out to design a method of programming a computer to learn from its past experience. The computer would first solve a simple problem. Then, before it got to the next problem, it would look back on how the previous one went and learn from the experience. Did it, in retrospect, make the most efficient decisions the last time around? How did things work out? These may seem like trivial questions, except that such computers aren't just operating on numbers, they're controlling physical objects in the real world, where a whole gamut of complicating factors can emerge. Mistakes can cost lives and money.

After writing her thesis and getting her graduate degree, Estlin landed a job at the Jet Propulsion Laboratory in Pasadena, California. JPL is the center of some of the coolest robotics on the planet and off it (JPL designs the rovers for NASA missions). Estlin was thrilled to land a job there because spacecraft are perhaps the most autonomous machines in the world.

A Mars rover, as you can imagine, has to be really resourceful. Once it is packaged up in the cargo bay of a rocket, it doesn't arrive at the Red Planet for many months. Once there, it operates its entire life without the possibility of any physical intervention. No technician is going to be able to swap out a bad chip or tighten a loose screw or help the rover get out of a pothole. And the computers that control Mars rovers aren't fancy. Any piece of equipment on a trip to Mars has to be "hardened" against the intense radiation of the sun and outer space that bombards any object that leaves the protective confines of Earth. To keep a stray cosmic ray from frying a computer chip, you have to wrap it in cladding made of conducting metals, which makes things heavy. On a rocket ship on a 35-million-mile journey, every ounce is precious. So the capabilities of the computers that run Mars rovers are necessarily restricted—probably equivalent to those of an Apple II, circa 1980.

Despite that restriction, however, Estlin and her colleagues at JPL are able to get an extraordinary amount of performance out of these machines, mainly because programming techniques have gotten so clever. And they're continually improving the software. Though technicians can't get a lug wrench on a Mars rover, JPL engineers can beam up software upgrades. In March 2010, Estlin completed just such an upgrade to the MER Opportunity Rover, which has been plying the surface of the Red Planet since landing in 2004. The software upgrade allowed the rover to choose which images it beams back to Earth. The rover covers so much ground and it has such a limited ability to send the data back to Earth, that it can now look at the pictures and make a decision about which are potentially useful to scientists, and which can safely be discarded. It would be like a camera that knew how to look at your photographs and discard the ones in which your mother-in-law's eyes are half-closed or the top of your uncle's head was cut off or that you took by accident as you were putting your camera in your bag.

As good as the Mars rovers are, however, they are rather primitive compared to the latest techniques that Estlin and her JPL colleagues are working on. "A lot of what my group does is what I would call automated planning and scheduling systems," she says. These systems endow machines with the ability to think for themselves, to juggle new information, to adapt their plans to achieve a goal, or to change their goals as situations change. The new information "might include new science goals you want the spacecraft to achieve, or updates about how much fuel you have left or data storage you have left," says Estlin. The computer then considers the new data, and "automatically determines the command sequence to achieve your goals, while still keeping the spacecraft safe and making sure to obey resource constraints."

"For most missions, this is a very manual process," she says. "Usually humans are coming up with the actual command sequence and checking it themselves. Planning lets you do that more onboard, and you can do that at different levels. At the far end of autonomy you can have the [software] planner handle almost everything that comes up."

Estlin got a chance to demonstrate the usefulness of this technology a few years ago, when she and her teammates at JPL designed a software upgrade to NASA's Earth Observing-1 satellite, launched back in 2000. EO-1 has three imaging instruments onboard, which gather data on crops, ecosystems, and the environment. The new software gave EO-1 the capability to make myriad decisions on its own about how to go about gathering this data. EO-1 "is now able to compile user image requests and plan them out based on orbital location and priority, looking at several thousand possible scene combinations over a week's schedule. It then decides how to best combine scenes into a sequence of single, double or triple images and plots out the best plan of action," according to a JPL press release.[2] As a result of the upgrade, EO-1 now takes 40 percent more images than it did before.

And, she says, "We were able to significantly reduce the number of people we have on the ground watching the spacecraft."

Estlin's programs represent the best autonomous programming we're capable of today, and they run on the equivalent of an Apple II. Imagine what they could do on an Earthbound computer, in the wrong hands?

SCOTT BORG IS something of an intellectual butterfly. He studied economics at the University of Chicago but also dabbled in physical sciences and anthropology. Then he kicked around Europe for a while, writing and doing research in cultural history. For a time he worked in the theater as a "play doctor." He would tape-record the audience during a show and pinpoint where people were beginning to lose interest, and then suggest fixes. In the 1990s, he got a gig as a writer for two Harvard Business School professors, Adam Brandenburger and Barry Nabeluff, who were working on a book about game theory and business, *Co-opetition*, published in 1996. While completing that book, Borg decided to get back into economics. He had had enough of the writing life.

Businesses had spent fortunes in buying computers in the 1980s without much to show for it, but in the 1990s productivity began to rise. Economists talked about value creation—ways of quantifying the benefits of things like information technology. The Internet was starting to become a cultural phenomenon. Borg tried to look at the business scene as a cultural anthropologist would. The economics that Borg had learned at Chicago, he came to think, was "too constrained by the assumptions of the eighteenth century." Instead, he was attracted to the notion that physical things—factories, products, people—were less important than most economists thought. "What things are made of matters less than how they are structured, and individual physical causes matter less than the broader interfaces between systems." It was an Internet-centric view, with networks of networks growing organically and unplanned, with no central control. The notion wasn't just about computers, but had an analog in business—companies, for instance, were relating to other companies and to their customers in more complex ways.

Then came the 9/11 attacks. Suddenly, vulnerabilities in our society began to stand out in stark relief. The Internet presented a massive risk that had largely been glossed over. When the Internet was in its early days, a computer virus might cause some disruption among a handful of academics, who were pretty much the only ones who used it, but there was no potential for great economic harm. The more computers that hooked up to the Net, the more ways for them to rely on one another, and the more economic activity they supported—and the more the potential for mischief grew. In the aftermath of 9/11, Borg found himself at an economics conference and struck up a conversation with an economist who had been looking into technical security issues. They were talking about the notion of measuring the value created by computers and networks to companies and to the economy. Borg was describing his work on value creation when the economist turned to him and said, "Can you do that for value *destruction?*"

That was a turning point for Borg. "I started thinking about how to destroy value," he says, "and how you could quantify that."

That night at his hotel, Borg went to his laptop and googled *cyber attack*. He found nearly eight hundred pages, which in those days was a lot. But fewer than ten focused on truly dangerous cyber attacks and none discussed the consequences to the larger economy. "They were all really just about shutting down computer systems." They focused mainly on things such as how much a retailer might lose if its order-management system were unavailable, or the lost productivity if a company's e-mail was shut down.

But Borg didn't think that shutting things down was the worst that could happen. "If you really wanted to destroy value, you would hardly ever shut down computer systems," he says. "You would hijack that computer system to do much worse things. In particular, to cause physically defective outputs to do physical damage to critical infrastructures. To take all of those systems I'd been analyzing in previous years that we were using to run our factories and our production facilities and use them either to destroy those facilities or have those facilities turn out things that were harmful. If you were clever, you'd either do it in a way that would physically wreck everything at once, or you'd do it in a way that the harm you were doing was not immediately apparent."

Borg began looking at all the industries he had been analyzing so as to figure out how computers could add value and began instead to figure out what harm could come to these industries if those computers, and the Internet they depended on, weren't merely shut down, but were actually taken over by terrorists or a hostile state and used to do damage. He looked at such things as the electric power companies and utilities, agriculture and food manufacturers, nuclear-power plants and oil refineries, banks and trading exchanges, and just-in-time manufacturing.

"So I looked at all the things I'd been analyzing and asked myself, 'What is the maximum harm we can do by taking control of these systems?' So one of the other things that was apparent that I was thinking about is that if you really wanted to destroy value, you wouldn't write little viruses or worms, you'd write longer programs. So I thought why not write longer programs that would function autonomously, and you could tell them, 'Look for this, and if you find it, do this. And if you don't find it, look for this other thing, and do that. Programs that would take control of these industrial systems in order to do harm without anybody manually intervening, having in an old-fashioned way to be at the controls."

IN SEPTEMBER 2010, a computer user placed a call to a computer security company to complain that his computer was running slowly.[3] The problem sounded very much like what millions of people face all the time when their machine is infected with malware—software that invades a computer and acts for some surreptitious purpose. For most of us, the software is downloaded inadvertently from a website or arrives as an attachment to an e-mail message. It can do something relatively harmless, such as gather browsing preferences for marketing purposes. Or it can do something more nefarious, such as hijack your PC and use it, along with thousands of others, to send out spam or launch attacks on websites. Criminal and activist organizations use these hijacked PCs, or "botnets," to gather credit card numbers and other info.

But this wasn't some ordinary home user complaining about a PC that had gotten bogged down with pop-up ads and spyware. Nor was it the typical business user whose customer resources management system was on the blink. It was a military facility, with safeguards against viruses, and the machine itself was a specialized computer chip for industrial machines. It didn't play video games, it didn't surf the Web, it didn't process words, it didn't calculate 15 percent of a restaurant bill.

The facility was in Iran, and the security company was VirusBlokAda, in Belarus.

Had this been ordinary malware, the Belarus company would likely have noticed it when it analyzed the computer code it received from its Iranian client (whose identity has not been revealed). When the security folks looked at the machine, they found some odd behavior. The machine would periodically reboot itself—it would shut down and start up all over again, for no apparent reason. The security company had looked through the code—the programming—on the computer, and couldn't find anything wrong with it. Still, the troubles persisted. So they sought help. They called Ralph Langner.[4]

Langner runs his own security firm in Hamburg, Germany, specializing in industrial machines—the same kind of machines as that which got the malware. Langner had a look at the code, and nothing jumped out at him as unusual. So he set up a lab test. He took some of the same computers—they were controller chips used in industry to run and monitor physical machinery, such as motors and turbines and anything that moves—and began running tests. He got a few of these controller chips, loaded up the software, and let them run. He would send each chip the kind of signals that it might receive in a real environment, with an eye to seeing what the malware, which was surely embedded somewhere in the software, would do. What would get it to switch into action? And what would that action be? He let the programs run and sat back and watched. It was a bit like catching a wild animal and putting it in a cage and seeing what it does when you poke it or throw in a piece of red meat or give it a stuffed animal.

Langner tried hooking an infected controller up with a bunch of others. "We could see on the first day when we infected our lab environment, which includes several different [types of] controllers, that it was talking to them," he says. "It was very happy when it found one. Then it started to get more and more bizarre. There was a lot of traffic going on."

Then Langner found something curious: the malware lit up with activity when it made contact with a specific type of computer chip, one used only for industrial machines, and specifically for machines built by Siemens, the German manufacturer. Langner ran a series of tests in which he poked the malware with a variety of signals, to get a hint at what it was programmed to do, what other things would trigger it into activity. He wanted to see exactly what the program was looking for. A lot of industrial systems use these controllers, but which kind of machine was this malware seeking? Was it designed to look for, say, the machines that built auto parts? Or operate chemical plants?

Whatever the malware was doing, Langner found it to be more sophisticated than anything else he had ever seen. That realization made him think about the purpose of the malware, and who might be the author, and what the target was. It would have to have been written by more than one person because it required a depth and breadth of knowledge that no one person would have. It required uncommon programming skills—the kind of autonomous programming that Tara Estlin used at JPL. It required a familiarity with these specific kinds of controller chips made by Siemens. It also seemed to understand how many other similar chips worked, but it had a particular fondness for the Siemens chips. Was it using the other chips to find its way to Siemens chips? And if so, why? Designing such malware would have required a sophisticated expertise in security, because the malware was insidious in its ability to avoid detection, adept at navigating networks, hiding in plain sight and springing to action only when necessary to its purpose—or so it seemed, without knowing exactly what that purpose was.

What organization had great technical resources and a strong reason to send malware to a specific industrial facility for purposes of undermining it?

As Langner pondered this question, he went about his life, which included sitting each morning over his coffee and reading the newspaper. Almost every day, news appeared about the latest wrinkle in the standoff between Western nations and Iran over nuclear weapons. Containing Iran's weapons program had been an obsession of the United States, Israel, and European nations for years.

Iran has a nuclear power plant in Bushehr, which it claims is solely for the purpose of providing civilian nuclear power. Iran also has a program to purify, or enrich, uranium. Enrichment involves separating uranium 235 (an isotope with an atomic weight of 235) from uranium 238 (a heavier isotope, with an atomic weight of 238), both of which exist in the same ore.

Separating the two requires spinning the ore in a centrifuge until the heavier isotope, uranium 238, settles at the outer edge, at which point it's a simple matter to skim off the lighter uranium 235. Uranium 235 is also known as weapons-grade uranium because it is ideal for making bombs. Iran's uranium enrichment program is so vast, argue Western intelligence officials, that it couldn't possibly have any other purpose than to make weapons. Iran has thousands of centrifuges operating around the clock in an underground facility, unreachable by Israeli or American bombs.

A centrifuge is basically a motor with an arm attached. The motor turns, the arm spins around like the blade of a fan, and at the end of the arm goes the sample of uranium. After it spins for a while, technicians change the sample, separate out the 235 from the 238, and load up another sample. The motor that drives such a centrifuge is simple—just like any other motor, really, with a coil and a magnet and alternating current that drives the rotor. But it is a precision motor. Nuclear engineers know exactly how fast the arm must spin to get the best results. They need the centrifuge to make tiny adjustments to compensate for power fluctuations, and they want to be able to hook the centrifuges up to a network, so they can collect data and otherwise monitor what all its centrifuges are doing at any time. For this reason, each centrifuge has a computer attached to it—an industrial controller, in fact.

For the small group of engineers who, like Langner, specialize in industrial controllers, and who see each other at the same industry conferences every year, it was common knowledge that Iran's engineers preferred to use the Siemens chip—the same chip that seemed to light up the malware in Langner's lab.

One day, over coffee and newspapers, it occurred to Langner that the malware in his lab might be targeted at uranium centrifuges, and that its authors may have been some Western intelligence outfit—the CIA or Israel's Mossad.[5]

"It was a guess," he says, "but it was a very strong guess. That was the most sophisticated piece of malware in the history of computing, and it was one hundred percent directed, so it must have been directed at a target that was worth the effort."

In the meantime, the malware had acquired a name—Stuxnet—and it was beginning to come under some scrutiny. The security firm Symantec later did an analysis of the distribution of Stuxnet—how far the malware had spread. Stuxnet had spread throughout the world, but it was apparent from a glance at the data that Iran was the target. That nation had by far the highest concentration of Stuxnet cases—60 percent of about one hundred thousand controllers—with the rest scattered thinly around the world.[6]

Symantec's analysis also found that Stuxnet, for a time, had been communicating with "command and control" servers in Malaysia and Denmark, possibly to give whoever sent Stuxnet out into the world one last opportunity to issue orders or collect information, should circumstances warrant it. Although Stuxnet was designed to be highly autonomous, its owners perhaps wanted to hold out some way of getting messages back and forth as long as possible before the malware was compromised, disabled, or discovered. Then one day these servers disappeared. Not in the physical sense, of course; they were always just virtual, borrowed servers that are available for hire, through shell companies and such, making them impossible to trace.

IN THE SHORT history of malware, Stuxnet represents a watershed. It has about as much to do with the worms and viruses of the days gone by as *Homo sapiens* has with the first amphibian that crawled out of the ocean. In the world of cyber warfare, worms and viruses are grunts, but Stuxnet is Special Ops. To appreciate its capabilities, consider how tough its assignment was.

The origins of Stuxnet are obscure, but it is almost certainly a new kind of cyberspy. Jason Bourne did not dress in black, cut through a chain-link fence, worm his way through the ventilation ducts, and stick his thumb drive into the uranium centrifuges. There was no need for a Jason Bourne because Stuxnet did the work for him. Stuxnet's creators gave their program the best training they knew how to give; they stocked it with plenty of detailed technical knowledge that would come in handy for whatever situation it could conceivably find itself in; they had rendezvous procedures and communication codes and so forth, so the software could report back to headquarters. But the main goal was to make Stuxnet autonomous. Like Bourne, it was built to survive and carry out its mission even if it found itself cut off.

At some point, Stuxnet was released from the confines of the lab into the wild, and it had to rely on its smarts and resourcefulness to reach its target: the nuclear facility at Natanz, where the centrifuges are housed.

If the computers at Natanz were ordinary PCs hooked up to the Internet and run by people who wouldn't know a virus from a worm, that might be easy. All you'd have to do is send an e-mail message with an alluring subject header, get it past the spam filters, and get some hapless operator to clink on the link. But Natanz is a military-hardened facility. The computers that operate the centrifuges are "air gapped"—meaning they are kept physically separate from the other computers at the facility, the ones that do the mundane things such as word processing and payrolls and are connected via the Internet to the outside world.[7] It wasn't enough to write the typical kind of virus and hope that somebody clicked on a bad link or downloaded an attachment to a spam message. Stuxnet had to somehow jump the gap. How do you get it to do that?

Nobody knew exactly how Stuxnet made its way from the lab where it was developed to the controllers of Iran's centrifuges. (Well, someone knows, but they're not telling.) But Langner, Borg, and other experts have speculated how. Stuxnet probably left the lab loaded onto one of those "thumb" drives that plug into the USB port that almost all computers have. The world of nuclear power engineers is a small club, so it would have been a simple matter to plant one or more of these USB drives on a nuclear power technician or engineer on his or her way to an industry conference. To pick one at random, the Nuclear Energy Institute runs an annual security conference, which in 2011 was held at the Sheraton San Diego Hotel. According to its website, attendees include "security managers, supervisors, trainers, officers and other professionals involved in security at nuclear facilities."[8] Attendance is limited to people who hold the proper clearances.

It might be simpler still to put the malware on a less-than-secure network that a nuclear technician is likely to plug his laptop or flash drive into, perhaps to collect a work order, file an invoice, or carry out some other bit of routine business.

Remember, Stuxnet is the Jason Bourne of the malware world. It's not necessary to bring it close to Natanz. All you have to do is place it a few degrees of separation away, and the software takes care of the rest.

It would be little trouble for American or Israeli intelligence to send an engineer or two to a conference with a couple of thumb drives. Although Stuxnet is a big program by malware standards—about half a megabyte—it is tiny compared to the kind of memory we take for granted today. It would fit easily on a four-gigabyte thumb drive ($6.49 on Amazon) alongside a few PowerPoint presentations and technical papers and pictures from last summer's vacation in Yellowstone. The engineer steps up to the podium to make his presentation, plugs his thumb drive into the laptop provided at the hotel, and bingo—Stuxnet jumps to the laptop. It has been released into the wild.

You would expect a laptop to have security software that would check any program entering from a thumb drive. It would run a search for known viruses and worms and so forth. But remember, this is not just any virus. It is not on any security company's watch list and bears none of the attributes of any known viruses. It is designed so that even in the unlikely event a computer scientist looks at its naked code, he's unlikely to notice anything out of the ordinary. It is well camouflaged.

When Stuxnet gets into the laptop, it hunkers down and waits. It doesn't want to attract attention. A mere half a megabyte, the size of an ordinary e-mail attachment, goes unnoticed among the hundreds of big images or PowerPoint or video files, which can each run to many megabytes. Stuxnet then goes to sleep, until the next engineer—let's call him Bob—makes his presentation, plugs his thumb drive into the laptop his host has provided, shows his PowerPoint slides, answers questions, and, during the polite applause, removes his drive, newly infected with Stuxnet.

Eventually, perhaps even later that night in his hotel room, Bob plugs his thumb drive into his own laptop. Eventually, perhaps that very night, he logs into the intranet of his company, Nuclear Inc., enters his ID and password, and does a bit of work. Stuxnet senses an opportunity to spread and wakes up. It works its way among the computer's innards so that it gets included in whatever data Bob uploads from the laptop to the computer system, making sure it makes its way past whatever filters the computer system might throw up. Stuxnet, again, because it has no identity that security software recognizes, no telltale characteristics, makes it past the company's virus scanners and other safeguards, assesses the firm's network, and starts looking for new places to spread.

Pretty fancy stuff, but we still haven't gotten to what makes Stuxnet so special. It has the ability to adapt to its circumstances and make its way from computer to computer in a way that gets it closer and closer to its target: to find the Siemens controller chips that run the centrifuges at Natanz, infiltrate them, and destroy them, in a manner that doesn't implicate Israel or America or whoever its authors were, and that does not raise alarms until the job is complete.

Stuxnet didn't just spread willy-nilly from one computer to the next. It had a road map of sorts in its head that told it when it was getting closer and closer to its target. Or perhaps it's more accurate to say that it had a road map, but didn't quite know the route it was going to take, and that it was prepared to figure out, as it went along, which parts of the map were valid and which weren't. Was there a bridge down or a cul-de-sac where the map indicated a through street? The malware would, for instance, have known when it had come in contact with a nuclear facility; when it had arrived at a nuclear consulting firm that worked with the Siemens controllers (perhaps by perusing the company literature or test documents or something else on the system); it may have been able to read the itineraries of technicians, targeting the laptops of those who had traveled to Natanz or to another destination that brought the malware within striking distance—say, by way of Pakistan or Libya or North Korea. When the unsuspecting technician logged on to check his e-mail, Stuxnet would make sure to insinuate itself into his hard drive.

But how exactly did Stuxnet make it across the air gap to the uranium centrifuges at Natanz? Surely Iran's engineers, knowing that the entire Western world was striving to upend its uranium enrichment program, had elaborate security procedures in place to prevent even the possibility of malware jumping the gap. Security protocols of an air-gapped system surely prohibit the use of thumb drives—especially ones that have previously been used on computers on the outside—on their air-gapped machines? How did Stuxnet overcome this hurdle?

If the creators of Stuxnet know, they're not telling. But the answer probably has something to do with human fallibility. Eventually, via Bob and countless other intermediaries, Stuxnet probably found its way to the laptop of a technician who works (or used to work) at Natanz. That technician let his guard down and inadvertently uploaded Stuxnet to one of the centrifuges.

How could this have happened? Technicians are human, and no matter how strictly an organization requires compliance to security procedures, someone is eventually going to violate them.[9] Security professionals, like many experts, sometimes think that the rules don't apply to them, that they're smart enough to avoid the worst without having to abide by the letter of the law. They might think that their laptops couldn't have any malware on them since they're so careful, and perhaps they are lulled by never having been a victim of any prior security breach. Industry insiders report that nuclear technicians often use laptops that run out-of-date software—old operating systems, perhaps, that aren't even compatible with the latest security software. One slipup is enough for a program such as Stuxnet, which has been proliferating steadily, ever vigilant to exploit the smallest chance of finding a way in. It could be that a technician who follows the rules 99 percent of the time slips up once and finds that he's plugged the wrong thumb drive into a secure system, or hooks up the wrong laptop, but is too embarrassed to report the lapse and figures chances are that no one will be the wiser.

One way or another, Stuxnet succeeded in infiltrating the Siemens controllers at the Natanz facility. What they did there is remarkable in two ways. One, they began to disrupt the centrifuges in a way that was subtle; technicians didn't notice until it was too late, and many centrifuges were completely destroyed. After a time, technicians at Natanz thought the machines were malfunctioning. A machine would break down, and they would have to take it off-line for repairs. Technicians would have a look, run their diagnostics, and not find anything wrong with them. When they turned the machines back on, they would seem to be running fine. When they monitored the operation of the centrifuges, to make sure they were working properly, everything appeared to be fine. The data on the monitor screen appeared normal. And the machine would break again.

How Stuxnet got the centrifuges to break down has to do with the physics of motors and the power that comes from being able to control what computers do. In uranium enrichment, one way you can destroy a centrifuge is by getting the rotor to turn a bit too fast, so that it pushes the extreme of what it was designed to do.[10] Any motor, whether it's in a car or a blender or a uranium centrifuge, is designed to operate safely, without damaging itself, within a range of revolutions per minute. But another RPM range will cause the shaft to begin to vibrate. If you run the machine at that speed long enough, the vibrations will eventually cause the shaft to break. That's what Stuxnet did.

Iran hasn't provided details of the disaster that befell its uranium enrichment plant after Stuxnet took over. Exploding centrifuges were probably a fairly common occurrence at Natanz, given that the centrifuges Iran uses are old IR-1s, a design going back several decades. (Pakistan's nuclear genius, A. Q. Khan, copied the design from Europe for Pakistan's nuclear program back in the 1970s and subsequently sold it to Libya, North Korea, and Iran.) Once in a while a rotor will break, parts will be flung at high speed, causing an explosion.

But it would be wrong to assume that Stuxnet caused fire and mayhem. For the most part, the malware made the centrifuges run just a bit harder, eventually causing them to break down. The manner in which Stuxnet broke the centrifuges would have been an important part of its design, because the longer Iranian technicians scratched their heads over what was going on, the more centrifuges Stuxnet could damage. "Obviously the attack was designed to give the operators the feeling of 'Oh no, several centrifuges have exploded, we should do our jobs better,'" says Scott Borg.

As the centrifuges began to break, managers at Natanz would have had conniptions: why are technicians not doing their jobs? The hapless engineers and technicians wouldn't be able to quickly surmise what was going on. Some of them may have been fired for incompetence. You can imagine the pressure on them, and their bosses, and the consternation in the ranks when the defense minister had to go to Tehran and tell President Ahmadinejad that the uranium enrichment program was falling behind schedule.

Why wouldn't engineers and technicians at Natanz have noticed that their machines were not operating as they were designed? Stuxnet may have misled them. Stuxnet was quietly destroying the machines while Iranian engineers stared at monitor screens that displayed false data indicating that all was as it should be. Before initiating its attack, Stuxnet had apparently recorded the data streams that went to the monitors, showing all systems to be normal, then replayed that data on the Iranian monitor screens. It could have been right out of a heist movie, with bank robbers feeding a video loop into the surveillance cameras showing happy customers lining up at bank tellers to deposit their paychecks, when in reality tellers are emptying the safe at gunpoint.

Eventually, Iran's engineers found the Stuxnet program and suspected that it was some kind of malware, but by then it was too late. Iranian engineers may have thought they were running the enrichment operation at Natanz, but in fact months earlier they had lost control to Stuxnet.

Ralph Langner made this point when he completed his analysis and reported back to the Belarus company that the Iranian machines had in fact been infected with an incredible new type of malware.

That's the kind of news you don't want to deliver in person.

STUXNET IS A remarkable accomplishment in the history of espionage, but the implications go far beyond cyber warfare to the very foundations of our increasingly technological civilization. We are now utterly dependent on computers and the Internet, and every day that dependence grows.

Because of people like Tara Estlin, who are bringing more and more autonomy to machines, the roots of digital technology are becoming intertwined with everything we do, and our well-being is tied up with it in ways that we don't even fully understand.

Even in the narrow mission of Stuxnet, the damage this program was able to inflict on Iran's nuclear weapons program has gone underappreciated. Even once Iran found out about the software, eradicating it was no simple task. The Iranians had to figure out a way to find the program and remove it, which would have meant they had to take everything off-line for weeks or months. In addition to cleaning up everything in Natanz, they would have had to make sure that other copies of the program weren't anywhere in the computers of any of their contractors or technicians offsite, because if one infected system is left, soon the entire operation will be infected once again. It might take Iran a year to sweep its operations clean of Stuxnet. Distributing Stuxnet in this way, throughout the industry, may have been part of the mission—to stall the Iranian program as much as possible.

We know that Stuxnet did not succeed in remaining undetected. Whether this is a failure on the part of its designs or was intentional is debated in cyber-security circles. Perhaps Stuxnet's designers were being cautious. Perhaps the malware's primary mission—taking out the centrifuges at Natanz—was so important that the planners didn't want to jeopardize the mission by giving the malware the additional task of covering its tracks. That might have slowed down the malware's progress from one computer to the next, reducing slightly the chances of success. Had Stuxnet been told to delete itself after a few months, sort of like the self-destructing tape in *Mission: Impossible*, leaving only the copies that had already moved on, then it would have been much more difficult to find. But at any one time there would have been fewer copies out there looking for a way to Natanz.

Stuxnet, in the end, had the destructive force of a military airstrike. It took out Iran's nuclear weapons program in the sense that the Israeli air attack took out Saddam Hussein's nuclear plant in 1981. It used deception in a new way. We've seen phishing attacks, in which a website or e-mail gets you to give up information by deceiving you into thinking you are filling out a legitimate form or logging on to your bank account. We've seen websites and e-mail messages that try to trick us into thinking they're from a friend or a legitimate business we deal with. But we haven't seen much software that can carry out a deliberate deception on the scale of Stuxnet. It's most likely the first of a new generation of software that doesn't just take things out, but inserts itself into the works surreptitiously and takes control.

The notion of machines taking over is a staple of science fiction. In the *Terminator* movies, machines develop their own notions of purpose and take control from the humans. The future isn't likely to be quite so clear-cut. At the moment, it doesn't look as if machines are going to gain self-awareness and start making the decisions for us. But we are already greatly dependent on digital technology, and the world is growing more and more interconnected into a vast digital machine. Software runs this machine, and if something goes wrong with the software, the consequences can be devastating. It doesn't take any great leap into science fiction to understand this. All we have to do is think about something as basic as our electrical grid, and what might happen if the software that runs it turned against us, just as Stuxnet turned Iran's own machines against them.

WINN SCHWARTAU ANTICIPATED the destructive potential of cyber warfare back in the 1980s, and wrote a novel about it, *Terminal Compromise*, which was published in 1991.[11] The novel's villain is a Japanese industrialist who lost his family in the Hiroshima bombing and is bent on revenge. He hires a young mathematical genius from the United States who, having been snubbed by the National Security Agency, is willing to sell his services to the highest bidder. (The math genius also had an insatiable appetite for sex, but that isn't germane to our purposes.) The result is a computer virus that creates considerable chaos. The computer systems that transfer money between banks go kablooey, and so do the ones that process airline reservations and support air traffic controllers. The Internet, of course, wasn't even a factor in those days.

In October 2003, security and military experts ran a simulation of a cyber attack on the United States. The idea was to figure out possible scenarios that security strategists should be worrying about. Known as LiveWire, the exercise looked at possible consequences of an attack on the United States. The study concluded that an attack would be "capable of inflicting significant material, financial and psychological damage through denial of service and related attacks that inflict a disproportionate toll on our economic welfare and our standard of living, in some cases."[12]

LiveWire may have been a turning point in how people thought about cyber attacks. Before then, they were generally thought of in terms of what would happen if the Internet went down—how would an outage effect commerce, the military, and so forth—or in terms of what secrets or confidential information could be stolen. But as Stuxnet would later show, shutting things down might not be the worst that could happen. It would be far worse if malware could take control of key machines and keep them running apparently normally, but be carrying out an elaborate deception, with the idea of inflicting damage that we can hardly imagine. "Children shut things down," says Borg. "Grown-ups hijack them and do much worse."

In the truly worst case, if things started to go awry in a big way and it became clear that computer systems throughout our economy had been compromised, had indeed been taken over, to some nefarious purpose, individuals, companies, and governments would be unplugging their own machines as fast as they could, to keep the damage from getting worse. One LiveWire scenario—an attack on financial systems—had hackers shut down ATMs, render credit cards useless, and interfere with check clearing and automatic deposits of payments. The group did a financial-sector analysis. What should the government do in such a circumstance to help soften the crisis? "We send in truckloads of money," said one of the attendees. He was being only partly facetious.

In the event that the financial system shut down completely, it might be wise to start handing out money to anybody who showed evidence of a checking or savings account, because at that point whatever was lost in fraud would be a drop in the ocean compared to the need to keep the economy moving.

An economy is complex, and it's hard to take it down completely—hard, but not impossible. One way to do it is to work on the notion of substitutes. If you take out ATMs, people might write checks. If you also destroy check-clearing systems, people might use credit cards. If you take those out, too . . . well, then things start to fall apart. "In an attack, if you take out something, it has almost no effect," says Borg. "Then you take out the thing that substitutes for that, and it has almost no effect. Until you take out enough things so there's no longer adequate capacity to substitute, and then you get really bad effects. If you can take everything so nothing can substitute, woomp—you go from slowing things down to nightmare."

• • •

WHEN BORG STARTED to study the destruction that could be wreaked on society from malware, the first item on the list was the power industry. The electricity grid is a vast system of generators that drive turbines, which feed their electricity out through transmission lines. As we found during the blackout in the eastern United States in 2003, the grid can be pretty finicky. The blackout was kicked off not by any surge in electricity, but rather the sudden removal of one section of the grid, which caused an imbalance, which in turn sent shock waves through the system.[13] It's a bit like a traffic jam, in which a minor accident can cause massive delays by causing drivers passing by to slow down and rubberneck. A drop in electrical output from one part of the grid created an imbalance, which strained circuits up and down much of the East Coast, causing the system to collapse.

In Idaho, the Department of Energy has a lab that since 1949 has been conducting research on nuclear power and other matters related to electricity generation. A few years ago, officials at the Department of Homeland Security began conducting a test to determine what kind of damage might be done to an ordinary turbine generator in a cyber attack. In the wake of 9/11, experts began to warn officials that the U.S. grid was vulnerable to malware. The experiment, dubbed Aurora, wasn't classified, but it wasn't publicized either, and the results were shown to a select audience of experts in and out of government. CNN got a copy of the video and has since posted it on YouTube.

With Aurora, researchers took a replica of a generator, the type that dots the nation and provides power from coal or hydro generators and feeds it into the transmission wires. They infected the generator with malware that was designed to disrupt its normal functioning. What they wanted to know was how badly malware could damage the power grid. They found that it could do quite a lot.

A generator is a bit like a centrifuge, in that it's got a motor—a rotor wrapped with a coil that turns through a magnetic field, held in place by the stator. What turns the rotor is steam or water falling from a dam, but the movement is controlled by computers, to keep things humming and to maximize the efficiency of the machine. But these computers could be infected with malware that throws things off slightly, throwing the rotor off-balance. The Aurora malware did just that, causing the generator to shudder and begin to smoke, as the rotor began rubbing up against the stator. Eventually the rotor shred the generator, caused it to catch fire, bent the shaft. On the video posted by CNN, you can see the machine shudder and smoke.[14]

When the video came to light, officials at the Department of Homeland Security asked CNN to withhold some key data—presumably so as not to make it too easy for would-be terrorists by giving them a recipe. But then the whole issue died down as the news cycle ended and everybody kind of forgot about it. Meanwhile, out in the hills and valley and plains of America, power is fed into the grid from a relatively small number of big, and very vulnerable, generators.

In the past few years, the utility industry and the federal government have been falling over one another to get the grid wired to the Internet. President Obama made $4 billion in grants available for "smart grid" technology. A smart grid is, at bottom, an electricity grid filled with computers and wired to the Internet—it means that every part of the grid, from the meters in your basement to the wires that carry electricity down the street to the switching stations and transformers and power lines that feed into bigger power lines that run eventually to the coal and nuclear and oil and natural gas and nuclear plants around the country—all this is peppered with computer chips, which talk to one another over the network, sending meter readings and flow rates and voltages and so forth. All that is essential to a modern electrical grid. The idea is for the disparate parts of the grid to talk to one another to more finely coordinate the flow of electricity and keep track of how much power each resident uses at what times, to perhaps charge more for peak demand. A smart grid might even compensate for the kind of imbalances that caused the 2003 blackout in the Northeast.

The smart grid is also greatly expanding the vulnerability to malware.

But the grid's vulnerability began long before that. Twenty or so nations now have the wherewithal to design some kind of malware that could infect the power industry, spread from thumb drives or espionage, according to Scott Borg and other security experts. Experts fully expect that Russia, China, and other present and former adversaries have *already* installed some kind of malware into our current power grid, for use at some later date, should it become necessary. (Likewise, you'd have to expect that the United States has returned the favor. It may even be irresponsible for a defense organization to have failed to do so.) If such malware has already been planted, it means it wouldn't even require the Internet to cause damage. The malware might even be built into the very hardware of the electronics that runs generators and power plants. These days companies who sell computer chips in the United States usually farm out the manufacturing overseas.[15] They use powerful computers to design the circuits, then send the plans in electronic form to other firms, who in turn send them to fabrication plants in China and India. It would be a small matter for a state such as China, say, to intercept these plans and insert circuits into the chips that would enable their engineers to control them from afar, or to wait for instructions to arrive before taking some kind of action that would disable the power grid. For all we know, such chips could be an integral part of the nation's grid already.

An Al Qaedea–like terrorist group isn't likely to be able to take such steps. It takes some expertise to intercept the design of computer chips and tweak them in a way that fits some grand nefarious plan, then return them to the original customer without tipping your hand. Likewise, coming up with a Stuxnet-like malware wouldn't be easy for a terrorist group living out of caves along the Pakistan-Afghanistan border. Stuxnet was most likely designed by a handful of people—fewer than a dozen—but these people possessed an enormous amount of expertise. They would have needed extremely talented programmers expert in the particular controllers that Siemens used, and those of others in the industry. They would have needed someone with knowledge of the nuclear industry, and specifically Iran's nuclear industry and gas centrifuges. (They would have had to know, for instance, how the program could tell an Iranian nuclear environment—the constellation of equipment it might find itself surrounded by—from, say, Pakistan's or China's.) All told, it only amounts to a handful of people. But this particular expertise is not easy for a small, ragtag group of people to assemble, at least at the moment. As time goes by, however, the expertise for things like Stuxnet will begin to disseminate, and it will become easier for small, nonstate groups to put this kind of malware together.

So what would happen then? What does all this mean to the potential to destroy the power grid? And what then? We've weathered blackouts before, what's the big deal?

Well, the big deal is that we've never quite weathered an event like this one. It's difficult to underestimate the importance of the power grid to the way our society functions. But let's take a little look at how bad an attack on it could be.

IMAGINE A VIRUS with all the sophstication of Stuxnet, targeted to taking out as much of the power grid of the United States as possible. Let's call it GridKill.

Don't worry too much right now about who created GridKill or where it got its start. It could have been in any of twenty or so industrialized countries, or perhaps it's a rogue group of experts that broke off from these countries or were hired by some terrorist group. Right now it would be difficult to imagine its coming from anywhere but a big country with a healthy defense industry, but that could change in the next couple of years.

It may start with a thumb drive, or an e-mail attachment, or a direct plant in the old-fashioned spymaster sense: someone physically loading it on to someone else's laptop. Perhaps the program inserted something into a digital meter in the basement of a residential property, uses the Internet to spread throughout the utility industry, and waits for an opportunity to slip onto the laptop of a maintenance worker, who one day, in a moment of laziness or forgetfulness, connects it to the computers that control a power plant. Suffice to say that GridKill was conceived, set loose in the wild, and began to spread.

GridKill has many of the qualities of Stuxnet. It can spread without being detected. In months, it makes its way from computer to computer and eventually begins to jump the air gaps and insert itself into the controllers at the generating stations throughout the country. For the longest time, it does nothing but sit there, disguising itself as just another hunk of computer code in a sea of code. (Stuxnet did this extremely well: even if a computer programmer looked at its code, he would see something so similar to the thousands of other programs in the system that he would probably have overlooked it.)

GridKill, though, goes further. Unlike Stuxnet, GridKill covers its tracks: each copy of the program deletes itself after it moves on, leaving no trace. Had Stuxnet done this, it might never have been discovered, and it would certainly have been much tougher to surmise its origin. So GridKill propagates through the Internet, infecting machines on the grid, until it reaches its target: the controllers that run the thousands of generators in power plants throughout the United States. All traces of the program—the copies on thumb drives and in laptops and on the Internet—eventually disappear into the ether. GridKill insinuates itself into the U.S. energy grid. A copy resides on just about every generator. Invisible, malevolent.

The malware becomes a resident parasite, in a sense, sort of like the *Sacculina* parasite that infects crabs. The female *Sacculina* enters a crab's body through a chink in its hard exoskeleton and takes up residence in its belly, eventually growing into a knob on the outside of the crab. The parasite begins to take over the crab's life. The host crab no longer molts or grows because these things would take away from the energy the *Sacculina* requires to grow. When *Sacculina* releases its eggs, the crab helps disperse them in the water, as though they were the crab's own progeny. The crab looks for all the world as if it were behaving normally, but it's a hollowed-out shell of its former self, acting according to commands coming from another entity altogether. This is the U.S. grid under GridKill. Unlike *Sacculina*, however, GridKill has no effect on the energy requirement of the power grid or the industry's computer systems; it is "a configuration, not a physical object," says Borg. It doesn't grow in any physical sense, it occupies no space.

So what does the GridKill parasite do? For the moment, nothing. It waits and it watches. It gathers information about the operation of the generators, fine-tuning its ability to ultimately destroy them. It gathers intelligence. Remember, GridKill is intelligent and highly autonomous, like a Mars rover. It can go behind the lines for a long time and adapt its plans to fit its goal. It may even be able to pick up commands from a distant commander, like an Al Qaeda cell gleaning some code language from an Osama bin Laden tape. The code may give it a green light to attack, convey a time and a place, or set some other kind of trigger—a rainy day, perhaps.

THE BLACKOUT OF 2003 in the northeast United States came about when a power surge in Ontario caused the grid to wobble. Electricity sloshed around the grid, and a cascade of failures swept Ontario and six U.S. states, shutting down 265 power generators that supplied power to about 50 million people. (Notice how few generators were involved.) In some areas, backup power generators kicked in, but most areas were left dark for nearly twenty-four hours. Cities lost water pressure; Detroit issued a boil-water order out of concern of contamination. Landlines were overwhelmed and cell phone networks shut down for lack of backup power. Amtrak trains plying the Washington–New York–Boston corridor stopped in their tracks. Television stations and hospitals used their backup generators. In many cities, sewage leaked into rivers and eventually into the Atlantic Ocean due to a loss of power to pumps. Airports shut down all over the area, and trouble remained even after power came back on because electronic ticket information couldn't be accessed. Gas stations were unable to pump fuel; motorists drove their cars until their tanks were empty, then abandoned them on the highways. Oil refineries shut down, leading to a spike in gas prices for days after. Factories and stock markets closed.

Power returned to most of the region within twenty-four hours. By five

A.M., power had been restored to New York City, a bit more than twelve hours after the blackout started. The economic toll was estimated to run between $4.5 and $10 billion, and eleven people died.[16]

How would that compare with our day of reckoning with GridKill? Let's start at the Grand Coulee Dam in Washington State.

The operators in the control room at the Grand Coulee Dam have been watching pretty much the same monitor screens for decades now, and hardly anything much happens. They aren't expecting anything to happen. The accident at Three Mile Island got out of control, in part, because operators were not paying close attention to the dials in their control room,[17] and that was in a nuclear power plant in the 1970s, with a considerably newer, and potentially more dangerous, technology than that of the Grand Coulee Dam, whose six big generators were built between 1967–74. This is not to cast aspersions on the hardworking people who operate the Grand Coulee's generators, but rather to explore a hypothetical scenario. Let's say, for the sake of argument, that the operators are sitting around, playing cards or perhaps video games, when GridKill awakens from its slumber and starts to make things happen. The turbines are turning a little faster every few minutes, they're beginning to exceed tolerances. The changes in speed are pushing the hardware to the limit of its ability to handle physical stress, and beyond. All the while, even if the operators in the control room are watching the readouts, they don't notice anything, because GridKill is feeding them normal readings.

Eventually, however, one of the operators begins to hear something. It's hard to make out at first, above the usual roar of the falls and the steady rumble of the generators. But the steady rumble doesn't seem quite so steady. There are now faint high-pitched sounds. Where are they coming from? The operator looks again at the meters. All seems normal. "Forget it," he says to himself, "I'm imagining things." An hour or so later, however, his ears prick up at the unmistakable sound of metal against metal. The monitors, which measure all aspects of the turbines, from the speed of their rotation to their temperature, all indicate normal operation—things should be just humming along. The operators exchange quizzical looks, then someone bursts into the control room in a panic. Smoke is billowing up from two of the generators. No, three. As they all rush out the door, there's an explosion, then another, and another.

The Grand Coulee has six large generators, each the size of a house, at the base of the 550-foot-high wall where the turbines can catch the energy of the water as it falls. Generators work virtually the same way a motor works—including a motor that drives a uranium centrifuge: you've got coils of wire that move in the presence of nearby magnets. In the case of a motor, the energy from electricity is being used to move the coils about an axis; in a generator, the axis is turned by water or steam, which in turn moves the coils through the magnetic field, which in turn generates electricity. The controlling programmable-logic computers are similar, as are the physics of the two types of devices.

Stuxnet cleverly caused the uranium centrifuges to throw themselves off-balance, inflicting enough damage to set the Iranian nuclear industry back years. A similar piece of malware on the computers that control the generators at the base of the Grand Coulee Dam would also cause them to shake, rattle, and roll. But because they're many times bigger, they should shake with far more vehemence than those puny test generators of the Aurora project. The physics of rotors and stators would amplify the destructive power of a few spurious commands from computers, and the generators would explode spectacularly.

The Grand Coulee Dam was build to withstand a lot of stress. An explosion among the six generators simultaneously is truly a black swan event: highly unlikely, but not impossible. The designers of the dam could not have conceived, back in the 1930s when they were drawing up the blueprints, that a series of ones and zeros—tiny electrical blips—could create such an event. The dam wall would probably survive an explosion without crumbling. If the attack were timed to occur during a major storm, there would be no electricity to open and close the spillway gates. The dam might not collapse, but it would suffer damage. Water would cascade down the Columbia River valley, sweeping up houses and buildings for miles. It would make a terrific scene for television and video reports, like something out of a disaster movie.

A collapse of the dam is not necessary, of course. The real destructive power of GridKill would come to the power grid and, eventually, to the economy. Imagine the failure of the generators of the Grand Coulee Dam being repeated thousands of times in utilities throughout the country, as GridKill went into action.

Let's say, for argument's sake, that GridKill didn't quite take out the entire electrical power of the nation. Let's say it just knocked out the Northeast and the West Coast. Imagine a blackout similar to the 2003 failure, but this time, it's not a ripple in the grid current that causes circuit breakers to shut down. This time the generators in power plants around the country literally explode. Power goes down, and it's not going to be restored in the next twenty-four hours. How long would the power remain out? Many months—perhaps even years.

It seems incredible that this should be so, but the worldwide capacity to manufacture generator parts is limited. Generators generally last thirty years, sometimes fifty, so normally there's no big need to replace the ones that are already operating. The main demand for generators is in China, India, and other parts of rapidly developing Asia. That's where the manufacturers are—none is in the United States. Generators are huge pieces of equipment, mainly made out of steel, a specialty product. Even if the United States, in crisis mode, put full diplomatic pressure on supplier nations—even if the U.S. military invaded and took them over—the capacity to ramp up production would be severely limited. Worldwide production of generator turbines amounts to a few hundred per year. If the United States suddenly put in an order for a thousand, it would take many years to fulfill. And that's assuming there were no other customers.

"We did a lot of tabletop discussion in the electrical power industry to try to figure this out," says Scott Borg. "If you declared emergency powers, commandeered expertise and facilities and had people working around the clock, and you really knew what you were doing, how long would it take to produce new generators, replacement generators? We asked people in the industry. We couldn't figure out any way to get the average time to anything less than probably six months, and maybe considerably longer. That's being optimistic. So you could deprive very large portions of America of electricity for months at a time."

A few thousand lines of computer code—taking up the storage space of, say, a home video or mp3, but a highly autonomous program, at the level of sophistication of a Mars rover software—could in principle deprive a great portion of America of electricity for the better part of a year.

Such an event would drive the United States, arguably the most technologically advanced and economically powerful civilization the world has known, into almost unimaginable suffering—it would result in disease, starvation, and death on a vast scale.

Let's start with Borg's analysis of the economics of value destruction.

If you shut down electricity for three or four days, in economic terms it costs almost nothing. It would certainly cause a lot of inconvenience. Emergency services would suffer. Industries that deal with anything perishable would sustain big losses. Perishables include not just tomatoes and milk but such things as airplane seats, the business of financial brokerage houses, gambling, trades on the stock exchange—anything that, if postponed, loses its value. Each day a casino isn't open is a day of lost revenue. A grounded airplane cannot make up for days spent sitting on the runway. Emergency services would be disrupted—patients in hospitals would die for lack of medicines or power to medical equipment, or a lack of transport. Elderly or infirm people would be stuck in hot apartments without air-conditioning. But for most of the economy, it wouldn't make enough of an economic difference that you would see the effect by the end of the financial quarter. That's even accounting for things such as just-in-time delivery, where factories rely on reliable and prompt delivery of parts. Generally, businesses have enough inventory to cope with anything on the order of a long weekend. There's enough extra capacity so you can make up three or four days easily in a couple of months. If a factory is running at 90 percent capacity, that means you can make up three days of downtime in about a month.

Beyond a long weekend, however, things begin to get ugly. Backup electrical generators in hospitals and other places begin to fail because they run out of fuel. Businesses run out of inventory and extra capacity. Grocery stores run out of food. After eight to ten days, more than 72 percent of all economic activity, as measured by GDP, begins to shut down completely. The only kinds of economic activity that remain are things such as the value you get from occupying your house or your apartment. Though, of course, you're sitting in the dark, with no heat and no air-conditioning, perhaps without running water. So most of what we think of as economic activity would be gone.

"We had a tabletop exercise in critical infrastructure industries," says Borg. "We asked people how much business can you do without—and we went through the other critical infrastructures. What would you do without electricity, water, whatever? What do you do after this many days? What do you do after that many days? We also went and interviewed anybody that had been out of electrical power. We found a hospital that had been out of electricity for many days and found out what they did to cope. We assumed that everybody was very ingenious, that people would learn how to get gasoline out of underground tanks using bilge pumps from boats. We assumed that people would just do anything you can think of to keep things going. And we still couldn't figure out how you would keep most of the economy going after ten days. Mostly everybody, even if they keep going, is dependent on somebody who isn't going to be able to keep going. We did tree graphs of the economy, we did work-flow charts. We started looking at what happens if you break this link and that link—woops, these people are isolated, they don't have what they need to keep going."

Emergency supplies would quickly run out. Hospitals would be out of critical medicines. Diabetics would go without their insulin, heart attack victims would not have their defibrillators, sick people would have no place to go. Food deliveries would virtually cease (no gasoline for trucks and airplanes, trains would be down). People would start to die by the thousands, then by the tens of thousands, and then the millions. The loss in human life would quickly reach, and perhaps exceed, the worst of the Cold War nuclear-exchange scenarios.

THE ELECTRICAL GRID is the low-hanging fruit, but it's not the only way to create a nightmare with a few keystrokes. Disrupting the oil and gas industry would have almost the same destructive power. A gas pipeline is, after all, full of explosively flammable substances. Raising pressures in some spots near ignition sources could cause explosions and damage that wouldn't be easily repaired. Refineries could be targeted, disrupting gasoline supplies. Gasoline is behind most of the shipping in this country, and if goods cannot circulate, the effects cascade through the economy. People have to go without, and the deaths begin to pile up. It's that simple.

Taking out the banking system could cause almost as much destruction as taking out the power grid. If you shut down the whole banking system, you've shut down a large portion of the economy. Borg's calculations show that a severe, long-term disruption to the banking industry would deflate 59 percent of economic acvitity (compared to 72 percent for the grid). That estimate takes into account that many people will continue to do business even if money cannot change hands. Banking, though, is more highly fortified than the utility industry against attacks, but perhaps not sufficiently so against software as intelligent as a Mars rover's.

Shutting down telecommunications would be pretty bad, too. Or you could destroy the railway system by causing trains to crash in tunnels and on bridges, or by damaging the tracks so that trains could not pass. Barges may not seem like a terribly large part of the modern U.S. economy, but if you cause them to crash and jam in the major rivers, the consequences pile up. Anyone who has lived in a region where life revolves around a river—the Nile or the Ganges or the Yellow River or the Rhine or the Mississippi—knows how important rivers are to economic life. Coal to electrical facilities in the United States mostly goes by river—some of the biggest owners of barge fleets in the country also own utility companies. River transport is in turn dependent on water-level monitoring, navigation, signaling, and locks—all of which are in some way under computer control. Even the rivers are digital.

"We're talking about something that's ten times worse than what the Great Depression was like," says Borg. "Shutting down eighty percent of the economy for an extended period, we're talking about something that's bigger than all the damage done to Germany during WWII, all the damage done to Japan.

"You asked for the worst scenario," he says. "It's really bad. It's hard for people to imagine, to get their minds around how bad this could potentially be."

THE WORST THING, perhaps, about malware like our imaginary GridKill bug is that it is virtually impossible at the moment to take any kind of systematic step to ward this malware off. There is no fortress we can erect that will keep it away, and no reliable way to tell if any computer systems have already been infected.

A piece of malware is, after all, nothing more than a bit of computer code—a string of ones and zeros, like everything else that makes computers go. Computer code gives instructions to a computer—add these two numbers, compare them to this other number, store the result for reference later, and so forth. All the actions of a computer can be boiled down to simple operations like these. Malware, therefore, is nothing more than instructions to a computer that causes it to do something we don't want it to do.

Computer security companies are fine at taking steps to fight malware that they already know about. Conficker A, a bit of malware that has spread among millions of computers,[18] is a known quantity in the sense that programmers can identify it and take steps to neutralize it. (It is not a known quantity in the sense that its purpose is entirely opaque, and its source is also unknown.) So computer security firms can scan for Conficker A because they know what series of instructions it consists of.

Iran couldn't do that for Stuxnet because nobody knew what Stuxnet was. It had nothing in common with other worms and viruses that made it recognizable.

The trouble is that it's hard to find something when you don't know exactly what it is. There is no way to come up with a formula that would identify an unknown (and cleverly designed) piece of malware. Experts have certainly tried to do so. They've tried looking for the kinds of things that malware might need for breaking and entering a computer system, for instance—the cyber equivalent of a cat burglar's grappling hook for scaling a wall. But malware doesn't need anything special to break into a computer system.

You can't even find malware by knowing where it hides because it doesn't necessarily look like anything that shouldn't be there in the first place. And even if it did, a single piece of malware can break itself up into bits and spread itself all over a computer or even a network, so that even if you identified one of the pieces, you wouldn't necessarily be able to find the rest.

That leaves us with one strategy left: to identify a piece of malware by its effects. When something goes amiss, when a computer starts doing something you don't want it to do, then it's time to look at it to see if some malware is causing it to misbehave.

But what if the first thing the malware does is cause the power grid to blow up, or the banking system to fail, or missiles to turn around in the sky and fly back at us?

Experts have no good answer to this question.[19]

THE ADVENT OF intelligent rogue computer programs such as Stuxnet is only one of the many ways the field formerly known as artificial intelligence is making its way slowly and inexorably into every aspect of life. This is what happens with technology. It starts out as something for an elite corps of supernerds and gradually works its way to the masses, getting cheaper and more powerful. A computer once was a vast hunk of expensive hardware that only giant enterprises could afford; a computer industry executive once infamously said that the global market for computers could only ever be as large as about a dozen, a figure he arrived at by counting the number of institutions that could possibly afford to pay the millions of dollars these machines cost at the time. Their computational essence is now mass-produced for devices that we carry in our pockets.

Nobody can predict the future. All we can do is avoid a gross failure of imagination. So with that in mind, let's take a few computer technologies that exist today and draw a straight line into the future, to see where they could possibly lead. It's not that these things are going to happen, only that something like them very likely will, for better or worse.

A few years ago, a group of computer experts met under the auspices of the Association for the Advancement of Artificial Intelligence to consider the possibility that one day our machines will be smarter than we are, and might want to push us aside and take over the running of the planet.[20] It's a common theme in science fiction, but these scientists were trying to think about how to put this problem on more scientific footing. One of the scenarios they considered is if a robot was invented that could invent a smarter version of itself. Theoretically it would eventually surpass the intelligence of its maker, we humans, and what then? The idea wasn't to answer such questions, but to begin to think about the risks we're beginning to take, and to ask if there are ways of putting limits on what machines can do.

Artificial intelligence started out decades ago with the promise of general-purpose machines that could think and act like humans. These hopes were dashed, though, in part because the goal was too ambitious—human intelligence is just too subtle, too sophisticated, too poorly understood, to capture in a machine. It failed, too, in part because the hardware was too crude—computers in the sixties, seventies, and eighties were big but not powerful. Now they're tiny and quite powerful, and getting more so every year.

In the meantime, computer scientists have taken a divide-and-conquer approach to the problem of artificial intelligence. They've broken it up into bits and attacked each one separately. This had led to something of a renaissance in the field in the past decade or so. "Artificial intelligence is developing rapidly," says Tom Mitchell, a computer scientist at Carnegie Mellon University. "Ten years ago there wasn't any voice recognition." Progress in AI is along narrow slices of intelligence. Unlike biological systems, which develop in toto, an AI system is narrow—it's speech recognition, text reading, computer vision. The pieces come together in robots, which have sensors to take in what's going on in the real world, and the ability to move about and to effect physical change on the world. Increasingly, robots interact with people and their daily lives.

The notion of humanoid robots taking over the world is probably silly.

It is certainly silly when you think of robots in the literal sense, as mechanical creatures with arms and legs that walk around in the streets and sit at a desk in the office cubicle next to yours, competing with you for a promotion. But it becomes less silly when you abandon the literal notion of robots as humanoids, or something approximating humanoids. In the world we're now creating, you can think of robots as any artificial intelligence that connects somehow with the physical world. In this respect, Stuxnet was a kind of robot; instead of affecting the physical world through its arms and legs, it did so through the uranium centrifuges of Iran's nuclear program. Robotics, says Mitchell, is "the one exception to the narrow view of AI." A robot is a general-purpose tool made up of different components of narrowly built artificial intelligences.

The experts convening to discuss the possibility of robots taking over the world did not seriously believe that we were in any danger of falling to a machine rebellion anytime soon, but neither did they dismiss the idea. "I think the probability of robots taking over the world is low," says Mitchell. "But it's something to think about."

The first concern that engineers express about new technologies is inevitably privacy, and machine intelligence is no different. Mitchell's concerns are the privacy issues, but you can see that they rapidly become something else. It starts with your iPhone. It is, basically, a computer, and it carries an awful lot of information about you. It's got a camera, a microphone, a GPS that gives your location. The kind of information it collects is very telling about you and your habits. And the degree to which this information is collected and made available is only going to increase. Many policymakers and computer experts are thinking up ways of using the kind of data that cell phones collect to improve such things as traffic control and public health. If you're home with the flu, for instance, health officials could use your cell phone data to figure out who got within three feet of you in the past few days, when you were at the peak of contagiousness, and use that information to help contain the spread of infection, perhaps by contacting those people and informing them that they are about to be sick and are unwittingly at that moment spreading infection.

Having your phone provide such information to, say, the CDC may offend your sense of privacy or you may think it's worth whatever threat to privacy for the common good. Regardless, imagine what would happen if a computer virus promulgated by organized crime infected your phone and began to turn its capabilities of information-gathering to nefarious ends.

A sophisticated virus in your cell phone might be able to listen in on all your conversations. It would know your credit card numbers, it would intercept all your e-mails. Microsoft, Google, and other firms have already developed software that prioritizes e-mail messages by what you're most likely to be interested in. They can do "sentiment analysis" that scans e-mail messages and finds out how you feel about certain things—whether you think Obama is doing a good job, or how you feel about the government, and so forth. The software can read blogs and automatically tag people as leaning to the right or the left on the political spectrum. The software could gather information the way Gallup polls do, but you wouldn't have to ask people what they thought about certain subjects; the software would be able to tell just by analyzing their e-mails and blog posts.

Vast resources are now available on individuals from a multitude of sources. If you have a machine intelligence that can draw this information together, you've got a mind that embraces the Internet and can sift through it with great speed and pluck out what information it needs. As a storage device, the Internet dwarfs all others. The human brain contains the equivalent of about 3.5 quadrillion bytes of information; the Internet contains ten times that amount.

But what would a robot whose mind embraces the Internet be able to do with all that information? That becomes clear once you start to look at the narrow bits of artificial intelligence that are now emerging.

The ability to speak has improved by leaps and bounds in the past decade. The mechanical voice you hear when calling the phone company to inquire about a bill seems more annoying than potentially destructive, but only if you fail to imagine the day when the ability to program a machine to understand language and speak it is just a tool that you can buy at RadioShack. The voice speaking to you from your iPhone and fielding your queries started out as advanced technology in elite labs a few decades ago, and now it's part of a common experience. IBM's Watson computer—now the world champ on *Jeopardy!*—is a terrific example of the progress engineers are making in giving computers the ability to move at will among humans. When Ken Jennings lost on *Jeopardy!* to IBM's Watson, he declared, "I for one welcome our new computer overlords." As a joke, it may have been a bit close to home.

Where natural-language ability gets dangerous, potentially, is when it gets a bit more powerful, then seeps down to common usage and becomes a relatively inexpensive tool that just about anyone can use. It's not just the ability to listen to spoken commands, it's a matter of interpreting human intent and responding in a way that sounds, well, human. If machines can do that well, then it may get harder to tell them from real humans.

What's also being programmed into these machines is the ability to move in the world of humans. Scientists talk about giving robots the ability to make "ethical decisions," but it's not ethics in the sense we normally understand it—not so much deciding between good and evil as being fair in a human sense.[21] Let's say, for argument's sake, that you have a robot that controls what television programs a group of people in a nursing home watch. One person, call her Angela, wants to see the BBC news, while Bob wants to watch Oprah. What does the robot do? Deciding involves considering such things as how many times Angela and Bob have gotten her or his way in the past, whose turn it might be, what interpersonal dynamic is at work between Angela and Bob, how other members of the group regard them, and so forth. There is an infinite variety of factors, a big gray area. The robot will be successful to the extent that Bob, Angela, and the other residents feel that it's being fair.

A machine that can understand a human by spoken language and can also move easily in the world of humans could do a lot of other humanlike things.

You can imagine using computer technology to impersonate a human—perhaps even someone you know. The idea of a computer that can sense human feelings and come up with an appropriate response is a legitimate subject of research these days, and companies such as Google and Microsoft have a keen interest in it. Crude emotive software has already been used with autistic children to bring out hidden social skills. As scientists understand more about how to simulate human emotions, they may increase the ability of computers to pass themselves off as human.

When you consider this possibility, you can imagine the kind of disruption that could ensue in a terrorist plot to use computers to impersonate people. This type of identify theft goes well beyond what we know now. It's not hard to go from these kinds of identity-theft scenarios to one in which machines (or software, which is a type of machine) orchestrate vast disruptions to our economy. You can imagine the confusion that would reign if software began impersonating important people, handing out conflicting commands, causing markets to tumble and people to behave in odd ways. It adds a whole new dimension to the kind of damage that a Stuxnet-like bot could do to the economy.

In a few years, if its promulgators had all the tools of natural-language processing available to them, you could get a phone call presumably from your wife—synthesized, of course, but you wouldn't be able to tell.

NEWSPAPERS REPORT ALMOST daily about the modern military's use of drone robots, used to great effect in Iraq and Afghanistan. These machines do all sorts of useful things. When Gary Powers flew over the Soviet Union in his U-2 spy plane and was shot down and taken prisoner, he had been gathering data for use by the U.S. military. These days drones do this work so that pilots don't have to take such risks.[22] The drones fly inconspicuously into enemy territory and send back video surveillance. They stand over a target and provide data to guide an attack by missiles or troops on the ground, and sometimes they shoot their own missiles. They do all these things effectively because they can act autonomously—they can make a lot of little decisions on their own about the logistics of their own operations. A controller at Langley in Virginia doesn't have to tell a drone to compensate for a gust of wind, or even describe the terrain it's likely to see. The drone has maps in its memory, and it has software that makes adjustments to its aerodynamics.

Drones can make all sorts of little decisions, but they can also make big ones—life-or-death ones. They are (or soon will be) able to recognize faces and surmise identities of individuals, and they could choose whether to shoot. This is not to say they are allowed to do so at the moment; they are not. They do not make decisions about whom to kill, they don't pull the trigger without some human at the other end of a video screen giving the go-ahead. There is great ethical debate over exactly how close a machine should be allowed to go even in targeting flesh-and-blood humans. But this is not a debate over autonomy so much as a debate about procedure. If you tell Jason Bourne to wait for a command before firing a shot, he'll still be Jason Bourne, fully capable of taking the shot on his own. By the same token, military drones are still capable of targeting and taking out an enemy on their own.

The question is what to do when the technology that makes it possible to build a drone becomes commonplace, when it's easy enough for states or even individuals to acquire the capabilities now reserved for the U.S. military, and who don't have qualms about sending a drone assassin into a crowd or a city with carte blanche to act. You could imagine a group like Al Qaeda or Aum Shinrikyo or Hamas getting their hands on drones that could take out political targets in Washington, D.C., or New York City.

That may sound a bit far-fetched, and at the moment it would be, but it won't be for long.

A peek somewhat further down the road in drone research is even scarier. Scientists have implanted computer chips in the brains of beetles.[23] The chip is connected to the beetle's nervous system and sends tiny pulses of electricity that make the beetle turn left or right or fly up or down. The chips also have little radio receivers that put the beetles at the remote command of their researcher overlords. In the lab, they've gotten the beetles to zig and zag and do loop-de-loops. It's not an ominous technology at the moment, but it does give the future of drones a new twist. You could imagine a swarm of locusts that respond to digital control wreaking havoc on crops. You could imagine a swarm of bugs with surveillance cameras in their mandibles fanning out across the land in search of particular people that some remote military power wants to target. Mix some gene manipulation in there and you could envision some kind of venomous creature under remote command that can inflict a paralyzing or fatal dose of poison. And so on.

If this sounds ridiculous, consider that some scientists are beginning to worry about privacy implications once drones get into the hands of law enforcement officials. If drone technology keeps marching along, you can imagine a day when the cops have drones the size of bees they could send to your house to see if you're growing marijuana plants or running a crystal meth lab in your basement.[24] The first worry about a new technology that is publicly expressed is often concern for privacy. We don't tend to tell real horror stories in advance—nobody wants to scare people off new technologies that could be beneficial. But you could just as easily imagine how such mechanical bees could, if in the wrong hands, cause considerable disruption.

Robots that operate with a degree of autonomy are already finding their way into civilian life. The medical industry, particularly in Japan, is already introducing robots that assist in the care and monitoring of the elderly and people with disabilities. A Boeing 747 can take off and land on its own, if need be. Autopilot software is already that good.

These days, full autonomy for airplanes and even cars is within the realm of the possible. (It's prohibitively expensive, but you can count on costs coming down rapidly; they always do.) Commercial aircraft have GPS location, they have systems that control pitch and yaw and overall aerodynamics. Some have dedicated maps that give them the ability to locate places and objects with great precision, and they can track moving objects as well. Airplanes can't quite recognize faces on the ground because the resolution they get isn't quite up to speed, but that will come in time.

Eric Horvitz, a software engineer at Microsoft, drives a Lexus that can turn the steering wheel to guide itself into a parallel parking space. That's a first step toward automating all of driving. Having the car do all the driving would save the driver time—you could do other things, such as work or answer e-mail or stare out the window. Sebastian Thrun, a robotics expert at Stanford, and his colleagues have already built a car that they think can drive itself from San Francisco to Los Angeles without the intervention of a human driver, and they hope to perfect it over the next few years.

Automation undoubtedly is a good thing. More than forty thousand people are killed each year in automobile accidents. "My view is that none of these accidents should happen," says Horvitz. We already possess the means to drastically reduce such accidents with software that takes over if a collision is imminent and takes steps to avoid it, by, say, applying the brakes more quickly than a driver ever could. "We could cut them by a quarter or in half with some basic machine intelligence," says Horvitz.

Face recognition, another narrow AI capability, also comes with benefits. Finding Osama bin Laden would have been easier had military experts been able to get hold of a recent photograph of him, which they could have given to their computers in the same way as a bloodhound is given the scent of the criminal they're chasing.

Of course, a lot of recent photographs of most of us are out there on the Internet somewhere—on Facebook or Myspace or Twitter or Flickr or Google Plus. If some malevolent organization didn't want you around, they could use automated face recognition—the kind Facebook now uses to tag your friends—and locate you with a drone.

Giving over decision making to computers has a dark side. "As machines move into areas dominated by human decision making, there's the risk that they'll make poor decisions for us," says Thrun. If a computer can understand us; if it can fool us into thinking that it is a person, or even a loved one; if it can see everything we write and say and knows where we are at all times; if it can marshal the full resources of human recorded knowledge (or at least what's recorded in bits and bytes)—if it can do all these things, we have left ourselves vulnerable.

Looking further down the road, we can foresee a day when autonomous intelligence robots get so small that we can no longer see them with the human eye. There doesn't seem to be much of a limit to how small these machines can get—down to the size of DNA molecules. Scientists have been trying to get DNA to work as computers, and they've been taking steps that would be the precursor to creating tiny robots out of DNA, or some combination of DNA and other materials. These nanomachines are on the scale of one billionth of a meter, which is not a whole lot bigger than viruses.[25] (The term *nano* is used to cover a multitude of sins, from machines the size of cells to those the size of molecules.)

Ned Seeman, a chemist at NYU, has been working most of his adult life on building structures out of DNA. He's succeeded in making a lattice out of DNA that, rather than running as a twisted helix, branches out like a fence, so it can actually form a kind of scaffolding upon which it might be possible to add parts of a nanomachine. Seeman has succeeded in making structures in two dimensions and is working on extending that to three dimensions.

If Seeman and other researchers are able to fashion DNA into tiny robots, it opens the possibility that these machines can be self-replicating—that they would be able to build more of their own kind, replicating like some kind of life-form. Or perhaps eventually they would learn how to pool their resources, to cooperate with one another, each specializing in building some part of a greater whole, much as human embryonic stem cells multiply and eventually begin to differentiate themselves into the building blocks of organs, which form the complex organism we know as a human.

We may, in other words, be fashioning our own replacements. Before nanomachines pose a worry, scientists will have to solve the energy problem—how do you provide an energy source that's small enough for a machine that's merely the size of a molecule? It's not a trivial problem, and it may be a deal breaker.

Then there's Ray Kurzweil, the author of *The Singularity Is Near*, who believes that humans will one day soon move our intelligences into robot bodies. If or when that happens, will we have won, or lost?

The fly in the ointment of autonomous machines driven by information, whether genetic or silicon-based, is, once again, the computer virus. We have no defense, as we've seen, against malware. Viruses of the computer kind, as well as the biological kind, hold the key to our destruction.

One thing to do is to take the problem seriously—and that was the idea behind the meeting of AI experts. One of the ideas the experts discussed was avoiding the "Windows phenomenon," in which everybody uses the same software, making the world a sitting duck for virus designers. Even so, that's not likely to remove the problem of viruses entirely.

"It's impossible to get rid of them," says Thrun. "We've created the entire culture of viruses. We need to reduce our exposure to this risk."

How do we do that?

"I wish I could answer this question," Thrun says.

Ingenuity

In the early 1990s I was working in a rented room in Greenwich, outside of London. It was January, and I was chilled to the bone by the damp gray British winter. I had a blanket over my legs and a space heater the owner of the house, a friend of mine, had asked me to use "only to take the edge off." The room was piled high with books, so when I needed to get my blood flowing, I would get up from my desk and wander among them. One day I picked up a copy of *Infinite in All Directions*, a book adapted from Freeman Dyson's lectures. This eclectic little book, as eclectic as Dyson's career and mind, ranged from the origins of life to rocketry to nuclear winter. I stole the book from her pile and never returned it. I hope my friend didn't miss it.

The stories in Dyson's books drew on his extraordinary breadth and depth of knowledge of science and engineering, and yet they were delightfully imaginative.[1] Some were downright wacky, and some of the strangest ones described the author's exploits.

Dyson had an endless supply of new ideas, both big and small. The occasion of the moon landing got him thinking about how to make space exploration cheap enough so that the average family could undertake a trip to the moon. In this sense, he wrote, the goal for space exploration is similar to the notion of sailing ships back in the days of the European explorers. When it was cheap enough for an average family to scrimp and save and borrow enough for passage, Europeans settled the New World. So perhaps the first step to settling the moon and the rest of the solar system was producing a cheap two-seat space shuttle that an average person could afford.

Dyson was arguably at his most eloquent when discussing the distant future, when he had to speculate on where developments in science might one day lead to new technologies:

> Anybody who lives in New Jersey can easily watch a Monarch butterfly climbing into its cocoon and then afterwards climbing out again. It is an awe-inspiring sight. Sooner or later, probably fairly soon, we will understand how that is done. Somehow it is programmed into the DNA and we should soon learn how to do it. It is likely that, within the next twenty-five years, this technology will be fully understood and available for us to copy. So it is reasonable to think of the microspacecraft of the year 2010, not as a structure of metal and glass and silicon, but as a living creature, fed on Earth like a caterpillar, launched into space like a chrysalis, riding a laser beam into orbit, and metamorphosing itself in space like a butterfly. Once it is out there in space, it will sprout wings in the shape of solar sails, thus neatly solving the sail deployment problem. It will grow telescopic eyes to see where it is going, gossamer-fine antennae for receiving and transmitting radio signals, long springy legs for landing and walking on the smaller asteroids, chemical sensors for tasting the asteroidal minerals and the solar wind, electric-current-generating organs for orienting its wings in the interplanetary magnetic field, and a high quality brain enabling it to coordinate its activities, navigate to its destination, and report its observations back to Earth.
>
> I do not know whether we will actually have a space butterfly by the year 2010, but we shall have something equally new and strange, if only we turn our backs to the past and keep our eyes open for the opportunities which are beckoning us into the twenty-first century.[2]

It would be easy to ridicule these ideas with the perspective of time. The year 2010 has come and gone, and quasi-biological butterfly spaceships are not a reality (though scientists have developed the technology of space sails, which drift into deep space on the pressure of the sun's rays). One of the things I admired most about Dyson, and still do, is that he approached the promise of technology with a sense of excitement, as though it were a creative and humane act that we should embrace. Lately, Dyson has attracted many critics, particularly for a fanciful idea he had to use bioengineering to create "carbon-eating trees" to soak up excess carbon dioxide in the atmosphere. The notion of growing things, using some hyper form of biotechnology, is a big theme in much of Dyson's writings. He has also expounded on humans evolving into beings that can live in outer space, at home in the vacuum, on comets in the Kuiper Belt at the outer reaches of the solar systems.[3] Pretty fanciful, and certainly not the kind of things you'd propose seriously to a world that's desperate for answers.

What really has gotten people riled up is his contrarian view of climate change. He argues that the notion that climate change could have catastrophic effects is not well supported by the science, and in any case the question of what to do about the possibility—whether to undertake, for instance, costly measures to reduce emissions, which might divert resources from some other earthly need—is a question of values rather than science. He holds particular skepticism about computer models of climate, and thinks that scientists in general, and climate scientists in particular, tend to put too much faith in them.

Dyson has tangled in particular with James Hansen, accusing him of turning climate into "ideology." Hansen does not dispute this charge; he admits that he has come off his perch of scientific impartiality to become an advocate. He just believes the matter is too urgent to do otherwise. Hansen dismisses Dyson as ignorant of climate science. Dyson admits that Hansen is better informed, but says that he lacks perspective.

I am not eager to wade into this spat. (Nicholas Dawidoff wrote an excellent article about it in the *New York Times Magazine*.[4]) But it is interesting how swiftly minds can close to a fanciful idea when politics intrude. The debate over climate has gotten so emotional that we're in danger of throwing out good ideas with the bad. Dyson has had his share of bad ideas. In an essay, he once entertained the possibility of paranormal events. We can all see the obvious foolishness of sending a spaceship powered by nuclear bombs into Earth's orbit, which a project Dyson worked on in the 1950s would have required. Dyson admitted the team's collective naiveté on this score when he wrote an essay about it in the 1980s.

What matters more than the merits of any particular idea is the way in which conversations about climate and many other problems we face have been framed—as a war of religions. I'm not talking about creationists or people who don't even acknowledge climate risk. I'm referring to how important issues are being turned into choices between technology on the one hand and nature on the other. Writer Kenneth Brower said as much in an article about Dyson's climate views. "Environmentalism does indeed make a very satisfactory kind of religion," he wrote. "It is the faith in which I myself was brought up. Freeman Dyson does not have the religion. He has *another* religion." That religion is secular and technological.

The problem with turning things into such a choice is that it's not a very good way of filtering the good ideas from the bad. Carbon-absorbing synthetic trees may be a dumb idea. And although neither they nor space butterflies are going to help us avoid our worst fate, we're going to need the spirit with which these ideas were hatched to solve the problems we have created. Tossing aside technological optimism is not a realistic option. This doesn't mean technology is going to save us. We may still be doomed. But without it, we are surely doomed.

AS A THOUGHT experiment, what would happen if climate began to go haywire—that temperatures were rising faster than expected, glaciers were melting, the worst case scenarios were coming to pass. What if the world had already gotten it together to cut emissions of carbon dioxide and conserve energy and switch to alternative energy sources, but these measures didn't go far enough or didn't work fast enough to avoid the worst?

What if there were a technology that we could have in reserve to deploy in just such an emergency—a risky technology, to be sure, but one whose risks might be infinitely better to live with than the certainty of a climate catastrophe? It would be a kind of doomsday measure, an act of desperation, a life preserver to float on for a while should the Titanic hit an iceberg and sink. Would such a technology be worth developing? Would it be worthwhile doing some basic research that would answer the question: Could it work? And if we used it, what would the consequences be?

Such a technology has been proposed, of course. It's called by various names—geoengineering, or climate engineering. Currently, little research is being done on it because it has become a casualty of the religious wars.

David Keith, a climate scientist at Harvard, may have done more research than any other academic on the subject, at least recently. In 2010, he published an article in the *Proceedings of the National Academies of Sciences* on his research into how sulfur released into the upper atmosphere could exert a cooling effect on the planet. To put it simply, the idea is to block sunlight, cooling things down on the surface. The sun shines down on Earth, much of its energy reaches the surface and flows back up as heat, carried up by convection and a general stirring of the atmosphere, and makes its way back out into space. In the past century we've increased the amount of carbon dioxide, which holds back some of this energy, making the climate warmer. (The actual physics and chemistry are wickedly complicated.) The proposition of climate engineering is that we'd turn down the amount of the sun's energy that reaches the ground, bringing down temperatures.

That's where the sulfur comes in. When you put sulfur (in the form of sulfur dioxide) twenty miles or so into the upper atmosphere, it undergoes a series of changes that eventually yields particles called sulfate aerosols, which reflect sunlight, providing a kind of chemical sunshade. They stay in the air for about two years before they drift down to the ground or the sea. One of the most alluring things about the whole scheme is that it's not permanent. If you do nothing, the aerosols simply drop out of the sky. So even if it turns out to be a bad idea, all you have to do is wait and things will fix themselves. How can we possibly screw this one up? (There must be a way.)

We know that injecting sulfur into the stratosphere would cool the planet because it happened before, in 1991, when Mt. Pinatubo erupted and threw out 20 million tons of sulfur dioxide. It formed a slight haze that spread over the entire planet and caused a slight increase in beautiful sunsets in the year or two after the eruption. The planet cooled by a couple of degrees, enough to show clearly above the noise of daily and yearly fluctuations in temperature.

Keith isn't the only person to have thought of exploiting this phenomenon to engineer the atmosphere. The history goes back to efforts at cloud seeding in the 1940s, and was resurrected in the context of climate change by Nobel Laureate chemist Paul Crutzen, who wrote a paper about it in 2006. Keith, though, has become a kind of spokesman for geoengineering.

Keith and others have argued that it would be a good idea to find out precisely what effect sulfates would have on temperatures and climate, how much harm they might do to the oceans and the ozone layer, and how to engineer an aerosol sunshade—where to put the sulfur dioxide, what type of nozzle to spray it out of, and so forth. So if the worst happens—if climate starts to change in ways that create huge and urgent problems—we might have a tool that could head off the worst damage.

Keith has gotten caught in the political crossfire of climate change with these ideas. Advocates of cutting emissions don't like geoengineering because they fear it will provide a fake shortcut out of the climate dilemma. Cutting emissions is hard—it requires altering the fundamental chemistry of our main sources of energy, a basic driver of human activity. The world has had a hell of a time cutting carbon emissions. Efforts to come to international agreements to reduce carbon have failed. The Kyoto Protocol was dead in the water when President George W. Bush refused to sign it in 2001, the Copenhagen talks in 2009 were a disaster, and Durban in 2011 didn't amount to much, either. With the economy in the tank since 2008 and unemployment high, climate has dropped off the list of public concerns. And yet carbon still accumulates in the atmosphere. In 2010, emissions rose almost 6 percent, nearly as fast as college tuition.

Shooting a few sulfates into the air, by contrast, is dirt cheap—estimates run to something like $100 billion a year. When you put that sum against the cost of retrofitting coal plants, taxing carbon, inventing all sorts of new technologies, and making cars more efficient, it's a pittance.

And that's what makes the sulfate solution so attractive, and so dangerous. It's like crack—cheap and addictive. Every two years or so, you send up the planes to shoot their rockets into the stratosophere, spewing sulfur dioxide, which then works its chemical magic, and the world keeps its cool.

Such an intervention in the complex chemistry of the stratosphere is rife with potential ill effects. Sulfates, for instance, would exacerbate the tendency of oceans to become more acidic, which is already causing great damage. In general, the effects of aerosols are poorly understood. It could be that injecting tons of them would have side effects that we cannot at present foresee.

And what happens if you do it for a few years, as a quick fix? The few years drag into a decade or two, or three, and in the meantime industry keeps belching carbon dioxide into the atmosphere. So you keep replenishing the shade, sending the planes up and spraying their sulfur dioxide into the stratosphere. Carbon concentrations reach 500 ppm, 600 ppm, 800 ppm. But nobody feels the effects. It's like having a toothache and, rather than going to the dentist for the root canal you know you need, you take more and more painkillers. Eventually the decaying tooth will reassert itself into your consciousness, or you'll overdose. Either way, it won't end well.

What happens if carbon builds up to those heights and then, for whatever reason—war, plague, rebellion of the machines—the planes don't replenish the shade? Well, then the planet begins to warm with a vengeance, because all that carbon dioxide that had been building up in the atmosphere starts to make things really hot. And it would be a sudden shift, a jolt to the planetary system, so who knows what would happen? Storms, droughts, floods—who knows?

These risks could be considerable. But we might one day be so desperate that these risks seem vastly preferable to doing nothing. That possibility argues for more, not less, research so we can understand what we may be getting ourselves into. Where you stand on the research question, though, depends on whether you adhere to the technology religion or the nature religion.

Keith, of course, doesn't want this geoengineering disaster to happen. He is a reputable scientist at Harvard. I've met him several times and spoken to him at length. He is a wiry man who wears glasses and talks very quickly. He seems utterly sane and reasonable, nothing like Dr. Strangelove, the mad rocket scientist played by Peter Sellers in Stanley Kubrick's 1964 classic about the Cold War arms race.

Keith, naturally, worries about being seen as a modern-day Dr. Strangelove, and for good reason. The political right uses Keith's ideas to poke their enemies on the left, and the left dismisses the same ideas as technological hubris. Some conspiracy-theory websites present this kind of work as part of some CIA-led plot to use "cloud whitening" to control weather for military purposes.

Keith has at times felt personally threatened. "We once had to call the police because of a [local] guy who read about how I was a mass murderer on the Internet and thought he should do something to make the world safer," he says.

It's no wonder that geoengineering has become a political third rail for institutions that fund science. What program manager at NASA is going to stick his neck out to argue for sending a U2 high-altitude plane up with a hose and a nozzle to spray sulfates so that Dr. Strangelove can write another research paper on how to destroy the world?

Because the subject is so charged, Keith expends a lot of energy qualifying his position. He sprinkles his speech with phrases like, "Not that I'm saying we should do this," or "I'm only arguing that ignorance is not a good policy."

He explained at a conference in 2010:

> If you decouple the question of how much we're going to cut emissions from the geoengineering question—*which I know you can't, politically*—but if you did, and you assume that we actually are going to do what we need to cut emissions, we'll have one day where global carbon dioxide concentrations peak. I'm hoping that it's before 2050 and we'll have global celebration and after that we're rolling over the hill and concentrations are going down. Whatever that peak happens to be—*and that's a decision about how hard we cut emissions*—what seems likely—*but not certain*—is that

> we would have less environmental impact as a species if, leaving that peak, we also smoothly shaved off some solar radiation. Not enough to fully compensate, but some. So thinking about that as a temporary peak shaving—*it certainly is the only way I'm comfortable thinking about it*—but even then, we don't even know enough to say for sure if we can do that.

Keith takes pains to emphasize that he isn't an unabashed advocate for technology. He is clear about the danger of overly technological solutions to climate. "What if the pendulum swung the other way and we went overboard with the technology? And then we started to think about not just mitigation but planetary augmentation, not just to reduce the effects of climate change, but to make things better?" Say, to use geoengineering to improve crop yields. "I'm not advocating these things," he says.

Keith also talks about a "regret scenario," in which we find ourselves in a climate emergency but can't react fast enough because we didn't do what research we needed to do to make geoengineering a viable last-ditch option. "In the regret scenario, we didn't use the technologies we had available to us because we were too politically correct or too scared to touch them. In this scenario, we hopefully do reduce our emissions somewhat and look back in 2050 and think that we might have been able to prevent all that Arctic melting or changes in precipitation."

All this is tiptoeing around a notion that nobody really wants to think about: that a world of 10 billion people will not be a natural world, but one run and maintained by *Homo sapiens*, in which both action and inaction carries its own risks. "In the very long run, if humans don't wipe themselves out with a big war, we will live in an engineered planet, in the sense that it isn't natural anymore."

THE CHOICE WE face about whether or not to pursue this technology is often coached in terms of being for or against technology per se, as though we have to choose between technology and our humanity. This is usually a false choice.

The anti-vaccine movement is steeped in the notion that the body knows how to defend itself against viruses, that vaccines cause more harm than good, and that health officials and doctors can't be trusted. It's a willful ignorance on the part of people who should know better. For instance, the

San Francisco Bay area is one of the wealthier and better educated in the country, and yet it's become a hotbed of the anti-vaccine movement. In some schools, 40 percent of children are unvaccinated against whooping cough, creating the possibility of a public health crisis over a disease that was supposedly neutralized by a vaccine introduced in the 1940s.[5] The need for vaccines to fight disease should be obvious, but apparently it's not. When you look at the specter of bioterrorism—the possibility that some kind of biological agent will crop up somewhere unannounced—the need for robust public-health systems, including surveillance and rapid response, seems obvious.

You can feel the reluctance of some proponents of biotechnology to avoid the third rail of this antitechnology sentiment. Should their technologies get too much attention, there might be a public backlash. It's easy to see how the idea of engineers building their own organisms from libraries of genetic parts could turn a few people off, even if you call the product of this work a "bio-device" rather than a Frankenmicrobe. And yet synthetic biology has tremendous promise to do good things, too. The University of California at Berkeley engineer Jay Keasling is trying to build microbes that will manufacture artemisinin, a malaria drug that is currently difficult to manufacture.[6] Craig Venter's effort to build biofuel-excreting microbes has obvious benefits—it would remove carbon and fill our gas tanks in one stroke. It would be "carbon neutral" because the amount of carbon that came from the tailpipe would be the same as the amount the bacterium used to make the fuel in the first place.

A significant part of the environmentalist movement has at bottom the notion of taking the world back to a simpler time, when humans were closer to nature, before technology got us into this mess in the first place.

The author Bill McKibben argues that the climate has deteriorated to such an extent that there's no time left to wait for our energy economy to change to something greener. It would take decades, he says, to switch from a coal-and-oil economy to one based on windmills and solar panels (he doesn't much like nuclear power). By then, the atmosphere would have catastrophic concentrations of carbon, and it would be too late. In his book *Eaarth*, McKibben advocates "backing off"—essentially, making do with less. By this he means going further than increasing energy efficiency by raising efficiency standards for cars and using better light bulbs. He talks about cutting back on "big agriculture" and opting instead for more people to work on small organic farms to supply food locally.

These may be sensible things to do. McKibben isn't the only one who advocates them—the economist Lester Brown has talked about using efficiencies of local farming as a way of increasing energy efficiency, of forgoing the luxury of having your pick of tasteless fruits at any time of the year from anywhere in the world, and instead eating more seasonally, from farms nearby, cutting down on the need to truck and fly food around the world.

For McKibben, the need to reduce emissions is urgent. He is a leader of 350.org, a group that takes to heart James Hansen's assertion that if the atmosphere contains more than 350 parts per million for an extended time, it will cause grievous harm. Hansen may be right on that score. McKibben is correct that the kind of change that's needed will be slow in coming, even if the world were suddenly to heed Hansen's warnings.

Doing with less may be necessary, but it is not likely to be sufficient to solve our problems. It may not be any more realistic, especially in the short run, than counting on some technological quick fix.

Lowering atmospheric carbon to 350 ppm is a formidable challenge. Carbon concentrations already passed that point and are heading toward 400 ppm. Climate negotiations over the past few years have focused on 450 ppm, but even that goal is proving to be something of a pipedream, in light of the lack of progress to date. "It's an absurd number," Dan Shrag, the Harvard climate scientist, told me. "Getting there would require shutting down every coal plant in the United States and China in the next thirty years, and not building any more anywhere else."

As Shrag points out, coal is the big problem, from the standpoint of atmospheric carbon emissions. It's abundant and cheap not only in the United States but also in China, which makes it almost irresistible as a source of energy and difficult to argue against politically. And yet it's also an extremely dirty source of energy, not just in terms of carbon, but traditional pollutants like soot and sulfur as well.

Hansen, as we've noted, has come out against the use of coal, period—regardless of promises of making it "clean" with new and unproven technologies. Taking coal out of the energy equation is not going to be easy in the long run, and it will be impossible in the short run. Even if Hansen somehow held the White House and Congress in thrall, there's still China and Australia and the rest of the world to deal with. China, which surpassed the United States as the world's biggest carbon emitting nation, is working on clean technologies such as coal gasification—a way of turning coal into a gas that can be burned relatively cleanly.[7] But China is also building dirty coal plants at a fast clip. It will take years to perfect gasification technology and years more to introduce it at a big enough scale to make a difference.

As the world burns more coal, technologies for mitigating its worst effects run up against opposition.

Scientists have started looking seriously at technologies for "sequestering" carbon—capturing it from the effluent of coal plants, and rendering it in some stable way apart from the atmosphere, burying it in caverns below the ocean or turning it into some kind of inert substance. Sequestration has many technical problems to overcome as well as a big public-relations problem. Ground was broken on a coal plant called PurGen, in New Jersey, that would turn coal into hydrogen and sulfur and pump the leftover carbon dioxide into an underground sandstone deposit deep under the Atlantic Ocean. Environmentalists have opposed the plant mainly because it is too risky. They also don't trust the sequestration plan, arguing that, say, an earthquake might cause the carbon dioxide to leak from its underwater reservoir in the rocks and belch a tremendous amount of carbon into the air. The New Jersey Sierra Club (among others) opposed the project in part on this ground.

Carbon sequestration carries risks, to be sure. The United States pumps about 1.5 billion tons of carbon dioxide into the air every day. To keep up, we would have to sequester 20 million tons every day. It would only take a small leak to put much of that carbon back into the air in a few decades. On the other hand, does this mean technology should be discarded offhand? Dan Shrag doesn't think so. "The New Jersey Sierra Club is crazy," he told science writer Mike Lemonick. "This is nothing more than a scare tactic. It's no different from the irresponsible scare tactics right-wing think tanks use."[8]

"Let's be honest," Shrag told me. "It's a very big mountain to climb. Even if we are wildly successful beyond our best dreams, carbon might stabilize at 500 ppm. That may be intolerable. People may not be able to face that."

It's hard to see how we're going to find a risk-free path.

ENERGY IS A big part of the equation, but food is, too. So it might be helpful to think about how we would feed a world of 10 billion people.

If you were to create a new global agricultural system from whole cloth, it would not look like what currently exists. The rise of big centralized farms that depend on monocultures—single species of crops grown on a vast scale, all subject to the same blights and viruses—is not a well-thought-out arrangement. It just happened to spring into being through an accident of history. What we need is to redesign it, to rearrange it, and that's where things get tough.

Norman Borlaug's Green Revolution was a wonderful thing. It increased productivity dramatically, and it fed lots of people who might otherwise have starved. But it has its drawbacks. One is that it calls for enormous amounts of fossil fuels—it requires tremendous amounts of fertilizer—and lots of pesticides. And it involves much tilling of soil and replanting, which means that diesel-burning tractors ride up and down the landscape turning over soil, which causes it to release carbon into the atmosphere. It's a high-maintenance arrangement. The United Nations in 2005 called agriculture (and it was referring to the modern variety, post–Green Revolution) "the largest threat to biodiversity and ecosystem function of any single human activity."

Scientists, though, have begun to noodle ways of taking the big, centralized farm that we have today and bringing it a bit closer to what existed in the first farms ten thousand years ago. This is the kind of thing that would satisfy both proponents of nature and technology. One idea is to replace the annual crops with perennial ones which don't need replanting every year. Instead of planting annuals like corn, wheat, and soy each spring, farmers would cultivate perennial varieties that would grow for many years, throwing down deep roots in the soil. This kind of farming would involve less tilling and less fertilizer (at least after the first few years). And since breeders could design crops with roots that occupy different layers of soil, you could have several crops growing together on the same patch of land.

Perennial crops are only in their infancy, however. Biologists still have to come up with these new crop breeds. They are working on directly domesticating wild perennials such as wheat grass, which grows in abundance, to give it the right mix of traits—pest resistance, yield, and so forth—to make it a suitable crop. A more promising method is hybridizing perennials with the annuals we now rely upon—crossing, say, wheat with wheat grass to produce a low-growing perennial. Scientists are also screening breeds for specific "markers" in their DNA that will help identify desired traits. Finding the right mix of genes that make up hybrid plants is a complex and delicate business, and it will take time.

Even if a smart young crop scientist came up with a winner today, would it take root? Perhaps not. Perennial agriculture requires wholesale changes in how our farms and our entire agriculture industry works. It would require different kinds of fertilizer and different types of machinery for planting and harvesting. It would require a new generation of experts to advise the farmers on planting and maintenance and so forth. It is difficult to imagine any of this happening soon.

Perennial crops aren't the only potential solution. No-till agriculture has taken hold in the United States and is spreading through farming, especially in other countries.[9] It's a big improvement over conventional methods—it reduces carbon emissions from soils and is less energy intensive—but it's not a silver bullet. Quasi-organic farming methods and reducing the foods we consume out of season so they don't have to be shipped thousands of miles will require rearranging the economics of farming.

Small-scale agriculture may not be up to the task of feeding the world through the end of the century. Current projections have the planet's population peaking toward the end of the twenty-first century at about 10.1 billion people, which means that by the end of the century there will be the equivalent of two more Chinas to feed.[10] (China's current population of 1.4 billion people is expected to decline to just under a billion, mainly because of the nation's fertility policies.) These projections assume that we don't see some unforeseen decline, due, heaven forbid, to one or more of the grimmer scenarios outlined in previous chapters.

A great many of those new people will be located in Africa. Indeed, the higher-than-expected population growth in Africa is the major reason the United Nations revised its forecast upwards to 10 billion from a peak of roughly 9 billion in the middle of the century. The current population of Africa, about one billion, is now expected to grow to 3.6 billion by the end of the century. A continent that can barely support the billion people that currently live there will have to feed more than three times as many. More disturbing, much of this growth will come in some of the poorer countries, including many along the Sahel, which is already experiencing encroaching desertification from the Sahara to the north. Malawi's population of 15 million people could soar to 129 million. Nigeria's 162 million could soar to 730 million.

How Africa's agriculture will support such a population explosion is not easy to say, especially considering the vicissitudes of climate and disease. What's needed desperately now, and will be needed even more desperately in the coming decades, is a way of boosting the productivity of African farms. That's going to require a different type of technology than the kind of thing Norman Borlaug came up with, which applies mainly to irrigated crops. African farms generally rely on rain to water their crops. Where is this second green revolution supposed to come from?

Part of the problem in Africa is local politics. For the past forty or so years, land rights have been an impediment to large-scale commercial agriculture of the kind that revolutionized other parts of the world and led to huge increases in yields. In the absence of big commercial farming in Africa, research into new crops has been left largely to a network of government programs and NGOs. Researchers now work with farmers in Africa, in programs sponsored by the Rockefeller Foundation (which decades ago supported Borlaug's work), to investigate particular problems the farmers have and find ways of solving them. "They work on genetic improvements to crops, soil fertility, irrigation, climate change, which is increasingly important, and nutrition," says Gary Toenniessen, head of the agricultural development initiatives at Rockefeller. A big problem in agriculture in Africa, is that climates are fragmented; a crop type that works in one region doesn't necessarily work in others nearby. Research on small markets generally doesn't pay, so farmers often can't afford the products—seeds and so forth—that they need, and products they need won't bring a profit to commercial firms. Even rice and wheat, which are big cash crops, depend to a great extent on research that's supported by the public sector. For crops like cassava, the public sector is all there is.

The development of African food sources, however, is restricted to a large extent by a ban on the use of genetically modified crops. This stems to a large degree from a similar ban in Europe, which has lifted only recently after more than a decade. While GM crops aren't a panacea, they are yet another tool to bring Africa crops that could help the continent cope with drought and the spread of rust and other ills.

Golden rice, a GM crop engineered to provide beta-carotene, has had a tough time winning acceptance. It was designed to benefit people in poor rural areas who have vitamin A deficiencies but can't afford supplements or fortified grains and who may be beyond the reach of NGO programs to provide these things. Regulation of golden rice, says Toenniessen, is so onerous that only the biggest corporations can afford to jump through all the hoops necessary to bring a new crop to market.

Many of the technologies for addressing the food problem may not be all that appetizing, (though perhaps no less appealing than a tour of a factory livestock farm or, for that matter, the kitchen of some restaurants). The urgency of making systemic changes in our energy economy also argues for the need for new technologies that are disruptive—that can quickly be mobilized to have an impact on the efficiency of agriculture. What this would be is anybody's guess. (Here's where you need those creative ideas.) Artificial meat may be one such idea. Feeding livestock with grains to supply meat to the United States, Europe, and Japan, and the growing middle class in China, India, and Brazil, is hugely energy intensive. If people aren't going to give up meat, making it in the lab would be a way of supplying it without the attendant environmental problems. Some scientists are working up synthetic meat in the labs—meat that can be grown like crops. We're not talking tofurkey, but actual animal muscle cells grown, without the animal, in cultures, and engineered to take on the texture and flavor of steak. Of course, a juicy synthetic steak is still a fantasy—right now the labs have made only blobs that look like mucus and have the consistency of oysters (but not the flavor). Scientists have taken pig embryos, extracted the "stem cells"—the mother of all cells, which possess the capability of turning into any kind of cell, nerves, bone, muscle and so forth—grown the stem cells in cultures, and gotten them to "differentiate" into muscle, the stuff of ham and roast beef. Rather than meat being shipped all around the world from large factory farms, you can imagine synthetic meat bioreactors on the outskirts of a city producing synthetic steak and ham and sausage, which in turn is sold in street markets to locavores.[11]

By the same token, it seems inevitable that humankind is going to have to take an active and enlightened role in maintaining the planet's wildlife, informed by science. We've already begun to do this, of course. Conservation scientists are making "genetic corridors" for big cats in South America and Southeast Asia that give them a chance of surviving among the burgeoning human population. For the most part, the cats do well hemmed into parkland set aside for their use, but to keep the gene pool healthy, once in a while a stray cat will wander for thousands of miles to mate with cats in distant lands—spreading its genes to their progeny. The need for jaguars and tigers to do this wasn't appreciated until relatively recently, when scientists analyzed the animals' genomes and drew parallels between their DNA with their mating behavior.

You could take the genomic revolution further by imagining reengineering wild animals that can live in the world as we've changed it. At the moment, frogs, bats, and honeybees have been dying off in recent years from infections that arise mysteriously—perhaps because of pesticides, perhaps due to complex effects of climate change, perhaps both. Scientists are already applying genetic manipulation to make food crops hardier in the face of drought and pests. Wild animals could similarly be engineered to survive the modern world. Would it be better to have a world in which honeybees contained a few artificial genes that made them resistant to colony collapse disorder? Or would it be better to have no honeybees at all?

The best argument against fiddling with the ecosystem and with the genetic makeup of wild things is that we don't possess the knowledge or the wisdom to do so without screwing things up. This is certainly true. Wisdom may be in short supply, but knowledge is coming fast, and with knowledge comes choices. As our knowledge of biology grows (or explodes), we will be faced with the choice of whether or not to act upon it.

There may be a third way—a way of going back to simpler times. Gretchen Daily, working with Paul Ehrlich at Stanford, sat down a few years ago to figure out how many people a sustainable world could comfortably carry. They considered how people live now, the planet's resources, and the known behaviors of populations. "We tried to find a population size that would maximize options overall and hedge against disasters. On the basis of energy statistics and assuming more or less contemporary (largely Western-style) aspirations and technological capabilities, highly efficient energy systems and resource use, and a closing of the rich-poor gap, we came up with an optimal population of around 2 billion people—less than a third of today's."[12]

Two billion people—the population of the planet in the 1920s, when my father was born. Going from 7 billion to 2 billion is quite an adjustment. If this is the path, let us hope we move down it slowly and by choice, rather than quickly, by imposition.

HOW LIKELY IS it that our worst fate will come to pass? The easy answer is that nobody knows. But we can hazard some guesses.

Most people I talk to believe that biology—viruses in particular—poses the most immediate threat. Scientists have worried at least since 1997 that bird flu could make the species jump to humans—that the H5N1 virus could mutate into a form that kills people with great efficiency and also spreads from one person to another quickly and easily. The creation of killer bird-flu viruses in the fall of 2011 is good news in the sense that scientists beat nature to the punch. Doctors could potentially use this knowledge of scary bird-human flu viruses to create vaccines that preempt a deadly outbreak, should nature decide to spin her genetic roulette wheel in a way that goes badly for us.

Synthetic biology—the kind that our intrepid high school students at Gaston school got into in 2009—adds another dimension to what nature and bioweaponeers can do. It doesn't seem to pose any immediate threat. It's not likely, for instance, that some college students in a contest are going to mistakenly (or otherwise) produce some horrible pathogen that wipes out civilization (though it might be a good idea for a movie). But in general the dispersion of the knowledge and techniques of microbiology creates a very real long-term threat of abuse. This is true of old techniques like gene splicing as well as the new ones of synthetic biology. The Federal Bureau of Investigation has officials assigned to iGem, the MIT competition, to keep track of where it all is going. It makes sense that the people assigned to watch out for our safety keep current on techniques, and they most certainly will try to keep track of who has the knowledge to use them. Over time, this will become a more complicated task, as the number of knowledgeable scientists increases.

Practitioners of synthetic biology are sometimes skittish about publicity. They have reason to be shy. Their craft has serious potential to be misunderstood, should its dark potential attract too much attention. The fear, of course, is that there will be a public backlash against the craft itself, that politicians will call for overly restrictive regulations and oversight that crimp the ability of scientists to do their research.

Before that happens, it's important to understand the nature of the choice we face. We do not have the luxury of being able to choose between advancing the technology or stopping it in its tracks. It's simply not possible to rein in a technology that has already trickled down into U.S. high schools.

The better question is: What do we do to protect ourselves? Steven Block, no doe-eyed innocent, is opinionated on the subject. "No one is going to be able to put the genie back in the bottle, I think, because the new biotechnology is already too well-established and available," he says. "The only practical recourse is to blunt the impact of any novel threats. It will be vital for 'white' biology to stay a step or two ahead of the 'black' biology, so that emerging threats can be mitigated or otherwise managed.

"I see the future as a sort of war of escalation. It will take biotechnology to fight biotechnology."

The threat from machines, as outlined in chapter 6, is also an immediate threat, at least when it comes to the power grid, which is highly vulnerable to foul play. Its delicateness was plain during the blackout of 2003, when an imbalance in one part of the system sent ripples of power failures throughout the region. And Stuxnet shows one way in which such an attack could take place. Our saving grace here is that few nations have the know-how to carry out such an attack. If they have no reason to attack, they won't. (And keep in mind it's also highly likely that the United States has the means to retaliate.) Iran and North Korea are wild cards. We worry about their nukes, but perhaps we should be equally worried about their computer scientists.

Malware is infuriatingly difficult to fight in any systematic way. The trouble, of course, is that there is currently no foolproof way of recognizing truly novel malware or predicting what it's going to do. The bad option is to think that there's an easy fix—that if only government or security companies had the legal means to poke their fingers into anyone's data or software, they could somehow solve the problem. Giving government carte blanche to rummage through everything is not going to protect us from the worst malware attacks.

Beyond the grid, the threat of intelligent software is real. Industries like banking, pharmaceuticals, and so forth are more fortified in their security than the grid, but they have vulnerabilities. The prospect of natural language posing a big threat is also a few years off. The achievement of IBM's Watson computer in winning *Jeopardy!* is impressive, but that kind of natural language processing is not yet available at Radio Shack, and few people have the expertise to make use of it. That won't last forever, but it will last a few years at least. Neither would I bet on nanorobots—the problem of how to supply them with energy is a big obstacle.

How do we reconcile our reliance on the Internet, which is proving to be an unreliable beast? To find out, I asked the same minds that conceived the original Internet back in the 1970s and 1980s. Danny Hillis was a precocious genius who invented the Thinking Machine in the 1980s, one of the first computers to divide computational problems into parts that each fed to a different computer. Bill Joy is credited with inventing key aspects of the Unix operating system, which forms the basis of much of the Internet, when he was a grad student.

Back in the 1980s, when the Internet spanned a few hundred people who knew one another, trust was something of a commodity, and it became a basis for the decisions the group made in developing the technologies. When Hillis wanted to hook up a computer to the Internet, he would ask Bob Kahn, one of the Internet's founding fathers, at the Defense Advanced Research Projects Agency, for an IP address—the number that identifies each computer on the Internet. "Bob would take out an index card from his pocket, give you a number, and then cross it off his list," says Hillis. Now IP addresses are relegated automatically.

As we begin to gather more and more knowledge of the endlessly complex biological world around us, the artificial machine world we've built is becoming more and more like the biological one—complex and beyond our understanding.

One of Hillis's favorite examples is the "flash crash" of May 6, 2010. In the space of ten minutes, the Dow plummeted 10 percent, then climbed back up a few minutes later. Traders stared at their Bloombergs dumbfounded at the fall, and were equally dumbfounded (and greatly relieved) at the rebound. It took months of analysis to conclude that the crash was the result of high-speed automated trading. But does that really explain much? The truth of what happened is hidden in the algorithms of hundreds of machines that feed into the markets and interact with one another in ripples of cause and effect that nobody has a good handle on.

Recently Hillis and Joy have been thinking about the problem of the Internet and working on a way to make it more secure, a little bit less wild. They think one way out of this mess is to design a kind of Internet within the Internet—a network of networks that is like the current Internet in many ways, but which is designed from the ground up for a world in which not everybody who uses it is necessarily trustworthy. The trick is to design it in such a way that it can spread organically, the way the original Internet did—with people joining it because they want to. It would also have security built in, so that if big Internet Service Providers were taken out (as they were in Egypt during the recent revolution), or if undersea cables that carry bulk traffic were severed, the Internet would slow to a crawl. Instead, messages might pass from one device to another, jumping from cell phone to laptop to iPad, taking multiple paths to their destinations.

THE MOST VEXING problem of all is climate change. It is vexing because it is so fundamental—it strikes at how we supply civilization with energy, the driver of all things. It is vexing because we don't know how immediate the problem is. Will cascading climate flips happen far off in the future? Or have they already begun? Or are the theories just wrong? Without knowing, we can't say if it's still possible to do anything to reverse the changes we've already seen. And we certainly can't say what, if anything, is possible politically.

This brings us to extinction, which is the most difficult question of all. The more scientists study the sun and the atmosphere and the oceans, the more they realize that the planet is a mind-bogglingly complex system. When you add to that the intricate interrelationships of species, you've got a puzzle that can't be solved on a rainy Sunday afternoon. One thing is certain: If a mass extinction does come, it won't be a matter of losing a few polar bears or having to put up with an infestation of English Ivy in the backyard. It will mean the end of many species upon which we depend, and the end of us. Is such a thing underway, and if so, can it be stopped? Opinions are divided on this question. The evidence isn't in. But the outlook is worrisome.

The prescriptions are going to be familiar to anyone who reads *Scientific American* and the *New York Times*. We need to cut carbon emissions, and we also need disruptive technologies that somehow change the energy equation. We need to be vigilant of the biological threats, which means not shrinking from doing the research into scary subjects. It's important to recognize the magnitude of our problems, and not be squeamish about finding ways to address them. Humanity is a bold assertion, a derisive snort at nature. We've beaten the odds so far. To continue beating them will take every good idea.

Acknowledgments

This book wouldn't have happened without Ben Adams and his colleagues at Bloomsbury, including copy editor Steve Boldt and production editor Laura Phillips, who were supportive and tolerant; what a pleasant and professional bunch of people. I'd also like to thank my agents, Susan Rabiner and Sydelle Kramer, for invaluable feedback and support, which included many hours turning over ideas.

I want to thank all the people who took time out to explain what they do and how the world works—many of whom are named in this book, and many of whom aren't—but especially Jennifer Dunne, Doug Erwin, Gabi Neumann, Steven Block, Tim Lenton, and Marten Scheffer. Thanks, too, to the excellent journalists I have had the great fortune to work with: Mariette DiChristina at *Scientific American*, a terrific leader and friend who was always encouraging and patient; colleagues Ricki Rusting and Christi Keller, who watched my back, and Phil Yam, Robin Lloyd, and all the editors and staff at *Scientific American*, who helped more than they know by sharing their knowledge, creativity, and friendship. Fareed Zakaria encouraged me at *Newsweek International* to explore many of the subjects that ultimately became part of this book. David Freedman gave feedback and encouragement at every step of the way. Eve Conant and Monique Mugnier helped with fact-checking and research. Andy Nagorski provided advice and good cheer. Anne Nolan, Mary Carmichael, Geoff Cowley, and Billy Goodman read chapters and gave thoughtful feedback. Jeff Bartholet and Tim Folger gave their time and good judgment at crucial moments.

I also want to thank my friends and family for their support, despite my inconstancy and complaining. Thanks most of all to my family—my wonderful children, Sophie and Ben, who have inspired me, and my loving wife, Jude, who gave her support so generously, holding everything together, including, at times, me. I dedicate this book to her.

Notes

Chapter 1

1. Keith Bradsher and Lawrence K. Altman, "A War and a Mystery," *New York Times*, October 12, 2004.
2. Anne Underwood, "How Progress Makes Us Sick," *Newsweek*, May 5, 2003.
3. Pete Davies, "The Plague in Waiting," *Guardian*, August 7, 1999.
4. Bradsher and Altman, "A War and a Mystery."
5. Kumnuan Ungchusak et al., "Probably Person-to-Person Transmission of Avian Influenza A (H1N1)," *New England Journal of Medicine* 352 (2005), 333–40.
6. Donald G. McNeil Jr. and Denise Grady, "To Flu Experts, 'Pandemic' Confirms the Obvious," *New York Times*, June 12, 2009.
7. Fiona Macrae, "The 'False' Pandemic: Drug Firms Cash in on Scare over Swine Flu, Claims Euro Health Chief," *Daily Mail*, January 18, 2010.
8. Dolly Mascareñas, "Swine Flu's First Fatality: A Chronicle of Deaths Foretold," *Time*, April 30, 2009.
9. I have taken much of this account from Richard Besser's remarks on the 2009 pandemic flu at the Center for Biosecurity conference, which took place on March 5, 2010, in Washington, D.C.
10. John Kelly, *The Great Mortality* (New York: HarperCollins, 2005), xiv.
11. Gina Kolata, *Flu* (New York: Simon & Schuster, 1999), 20.

Chapter 2

1. American Association for the Advancement of Science, "Discoveries that Saved the Large Blue Butterfly Detailed," *ScienceDaily*, June 15, 2009.
2. Ed Yong, "How Research Saved the Large Blue Butterfly," Not Exactly Rocket Science, June 15, 2009, http://scienceblogs.com/notrocketscience/2009/06/how_research_saved_the_large_blue_butterfly.php
3. T. R. New, *Conservation Biology of Lycaenidae (Butterflies)* (World Conservation Union, 1993), 41.
4. David Derbyshire, "The Blues Are Back: Butterfly Species Reborn 30 Years After 'Extinction,'" *Daily Mail*, June 16, 2009.

5. Carl Zimmer, "How Many Species? A Study Says 8.7 Million, but It's Tricky," *New York Times*, August 23, 2011.
6. North American Butterfly Association, www.naba.org/qanda.html.
7. Jonathan Weiner, *The Beak of the Finch* (New York: Vintage, 1995).
8. American Association for the Advancement of Science, "Asteroid Killed Off the Dinosaurs, Says International Scientific Panel," *ScienceDaily*, March 4, 2010.
9. "12 Events That Will Change Everything," ScientificAmerican.com, May 21, 2010, www.scientificamerican.com/article.cfm?id=interactive-12-events.
10. Betsy Mason, "NASA Falling Short of Asteroid Detection Goals," Wired.com, August 12, 2009, http://www.wired.com/wiredscience/2009/08/neoreport/.
11. "The Catalystic 1991 Eruption of Mount Pinatubo, Philippines," U.S. Geological Survey Factsheet 112–97.
12. Much of the description in this chapter came from Douglas H. Erwin's *Extinction: How Life on Earth Nearly Ended 250 Million Years Ago* (Princeton: Princeton University Press, 2006).
13. E. O. Wilson, *The Future of Life* (New York: Knopf, 2002).
14. A mass extinction event is the product of many causes—a perfect storm of climate and biology and geology. In the Permian, many species were dying off. Some species of snails had been declining for millions of years prior to the vaporized coal basin. The seas had been receding, most likely because of some shift in the Earth's mantle, which caused an expansion of the crust—that would have been a big blow to many ecosystems on the continental shelves, having nothing to do with volcanoes, coal, or carbon. Yet, as Erwin puts it, "There is a really tight correlation between the extinction and the eruption of the Siberian flood basalts. And there's fairly good evidence that we had a greenhouse effect."

    Certainly other explanations are possible—meteor impacts or a critical mass of carbon dioxide in the ocean are two that have been mooted—but at the moment, the favorite is greenhouse gases. This is far from conclusive, but it is highly suggestive, and there are no other prime suspects. More likely, says Erwin, the greenhouse effect caused the so-called anoxia in the oceans. And it's hard to be sure when the actual volcanic flows triggered the carbon-methane release. So the bottom line is: all this is preliminary.
15. American Association for the Advancement of Science, "Oxygen-Free Early Oceans Likely Delayed Rise of Life on Planet," *ScienceDaily*, January 10, 2011.
16. Breck Parkman, an archaeologist at the California State Parks Department, studied rocks on California's North Coast that go about shoulder-high to an elephant. Indeed, when Parkman analyzed the rocks he found traces of minerals from mud that had insinuated themselves into irregularities in the rock surface over the years (see parks .ca.gov).
17. Ibid.
18. Jacquelyn L. Gill et al., "Pleistocene Megafaunal Collapse, Novel Plant Communities, and Enhanced Fire Regimes in North America," *Science* 326, no. 5956 (2009), 1100–03. Gill's work initially created a bit of a stir because it suggested the big herbivores had started dying off thirteen hundred years before the Clovis people were known to

have come to the Americas. But other researchers are independently pushing back that date. While scientists still have a lot of blanks to fill in, the overall picture that's emerging favors the hunting-to-extinction hypothesis.

19. Some fossil finds suggest that mammoths were also present in the Canadian arctic between 7,700 and 3,700 years ago, which could mean, incredibly, that they somehow survived and then went extinct only recently. But this doesn't change the fact that the first Americans had a terrific impact on the ecology of North America. (See James Haile et al., "Ancient DNA Reveals Late Survival of Mammoth and Horse in Interior Alaska," *Proceedings of the National Academy of Sciences* 106, no. 52 (2009), 22352–57.
20. Wilson, *The Future of Life*.

CHAPTER 3

1. Reprinted in the *Illinois Natural History Survey Bulletin* 15, no. 9, 537–50.
2. Marten Scheffer, *Critical Transitions in Nature and Society* (Princeton: Princeton University Press, 2009), 109–37.
3. Marten Scheffer et al., "Early Warning Signs for Critical Transitions," *Nature*, September 13, 2009.
4. James Hansen, *Storms of My Grandchildren* (New York: Bloomsbury, 2009), ch. 7.
5. NASA Venus Fact Sheet, http://nssdc.gsfc.nasa.gov/planetary/factsheet/venusfact.html.
6. Fred Pearce, *With Speed and Violence: Why Scientists Fear Tipping Points in Climate Change* (Boston: Beacon Press, 2007), 7.
7. Hansen, *Storms of My Grandchildren*, 116.
8. American Association for the Advancement of Science, "'Snowball Earth' Hypothesis Challenged," *ScienceDaily*, October 12, 2011.
9. Elizabeth Kolbert, "James Hansen Arrested," News Desk, June 24, 2009, http://www.newyorker.com/online/blogs/newsdesk/2009/06/elizabeth-kolbert-james-hansen-the-arrested-scientist.html.
10. Hansen, *Storms of My Grandchildren*, 236.
11. James Lovelock, "The Earth Is About to Catch a Morbid Fever that May Last as Long as 100,000 Years," *Independent*, January 16, 2006.
12. Scheffer, *Critical Transitions*, 150.
13. Nicolas Caillon et al., "Timing of CO_2 and Antarctic Temperature Changes Across Termination III," *Science* 299, no. 5613 (2003), 1728–31.
14. Philip Conkling et al., *The Fate of Greenland: Lessons from Abrupt Climate Change* (Cambridge: MIT Press, 2011), 101.
15. Ibid., 69.
16. Scheffer, *Critical Transitions*, 157–58.
17. Conkling, *The Fate of Greenland*.
18. Richard Alley, "Abrupt Climate Change," *Scientific American*, November 2004.
19. The nine tipping points detailed in this chapter are based on Timothy M. Lenton et al.'s "Tipping Elements in the Earth's Climate System," *Proceedings of the National Academy of Sciences* 105, no. 6 (2008), 1786–93, and on a conversation with the author.

20. For this discussion of the Sahara, I am indebted to Scheffer's *Critical Transitions*.
21. Rob Young and Orrin Pilkey, "How High Will Seas Rise? Get Ready for Seven Feet," Environment 360, January 14, 2010, http://e360.yale.edu/content/feature.msp?id=2230.
22. Richard S. J. Tol et al., "Adaptation to Five Metres of Sea Level Rise," *Journal of Risk Research* 9, no. 5 (2006), 467–82.
23. "Drought, Wildfires, Put Brazil under Environmental Emergency," Terra Daily, September 7, 2010, http://www.terradaily.com/reports/Drought_wildfires_put_Brazil_under_environmental_emergency_999.html.
24. Paul Jay, "The Beetle and the Damage Done," CBS News, April 23, 2008, http://www.cbc.ca/news/background/science/beetle.html.
25. C. Goldblatt et al., "The Great Oxidation at ~2.4 Ga as a Bistability in Atmospheric Oxygen Due to UV Shielding by Ozone," *Geophysical Research Abstracts* 8 (2006).
26. In this passage on Australia I have relied to a great extent on the Pacific Institute's report *The World's Water*, vol. 1, and in particular on the Matthew Heberger's chapter "Austalia's Millenium Droughts: Impacts and Responses," 110.
27. Frank R. Rijsberman, "Every Last Drop," *Boston Review*, September–October 2008.
28. D. Stehlin, "Managing Risk: Social Policy Responses in Times of Drought," *Disaster Response to Risk Management: Australia and National Drought Policy* (2005); Heberger, "Australia's Millennium Drought."
29. "Dust Storm in Australia," The Big Picture, September 23, 2009, http://www.boston.com/bigpicture/2009/09/dust_storm_in_australia.html.
30. Heberger, "Australia's Millennium Drought," 107.
31. Ibid., 27.
32. Peter H. Gleick, "China and Water," in *The World's Water, 2008–2009: The Biennial Report on Freshwater Resources* (Washington, D.C.: Island Press, 2009).
33. Ibid., 33.

CHAPTER 4

1. Herbert D. G. Maschner et al., "An Introduction to the Biocomplexity of Sanak Island, Western Gulf of Alaska," *Pacific Science* 63, no. 4 (2009), 673–709.
2. "A Run on the Banks," *E Magazine*, February 28, 2001.
3. Andrew C. Revkin, "Tracking the Imperiled Bluefin from Ocean to Sushi Platter," *New York Times*, May 3, 2005.
4. "Predator Loss," Save Our Seas Foundation, www.saveourseas.com/threats/predator loss.
5. Peter Kareiva, "Why Conservation Cannot Continue Ignoring Apex Species," Cool Green Science, July 20, 2011, http://blog.nature.org/2011/07/conservation-apex-species-predator-peter-kareiva/.
6. Ransom A. Myers et al., "Cascading Effects of the Loss of Apex Predatory Sharks from a Coastal Ocean," *Science* 315, no. 5820 (2007), 1846–50.
7. William J. Ripple and Robert L. Beschta, "Wolves, Elk, Willows, and Trophic Cascades in the Upper Gallatin Range of Southwestern Montana, USA," *Forest Ecology and Management* 200 (2004), 161–81.

8. Darryl Fears, "Study Links Fungus to Bat-Killing Disease," *Washington Post*, October 30, 2011.
9. Ker Than, "Deadly Frog Fungus Spreads in Virus-Like Waves," National Geographic News, April 1, 2008, http://news.nationalgeographic.com/news/2008/04/080401-frog-fungus.html.
10. David Biello, "Norman Borlaug: Wheat Breeder Who Averted Famine with a 'Green Revolution,'" ScientificAmerican.com, September 14, 2009, http://www.scientificamerican.com/blog/post.cfm?id=norman-borlaug-wheat-breeder-who-av-2009-09-14.
11. Gregg Easterbrook, "Forgotten Benefactor of Humanity," *Atlantic*, January 1997.
12. "Industrial Agriculture: Features and Policy," Union of Concerned Scientists report, www.ucsusa.org/food_and_agriculture/science_and_impacts/impacts_industrial_agriculture/industrial-agriculture-features.html.
13. Chris Fedor, "Rethinking the Sandwich: the Globalization of Wheat Rust," *Miller-McCune*, July 26, 2010.
14. Kate Melville, "Rust Never Sleeps," *ScienceNews*, September 25, 2010.

CHAPTER 5

1. Ann Finkbeiner, *The Jasons: The Secret History of Science's Postwar Elite* (New York: Viking, 2006).
2. Thomas J. Török et al., "A Large Community Outbreak of Salmonellosis Caused by Intentional Contamination of Restaurant Salad Bars," *Journal of the American Medical Association* 278, no. 5 (1997), 389–95.
3. Nicholas D. Kristof, "Japan Arrests Leader of Doomsday Cult in Subway Gas Attack," *New York Times*, May 21, 1995.
4. Two years after the retreat, Block made his insights in "Living Nightmares: Biological Threats Enabled by Molecular Biology," a chapter in *The New Terror: Facing the Threat of Biological and Chemical Weapons*, ed. Sidney D. Drell et al. (Stanford: Hoover Institution Press, 1999).
5. Ibid., 42–43.
6. Ibid.
7. Daniel G. Gibson, et al., "Creation of a Bacterial Cell Controlled by a Chemically Synthesized Genome," *Science*, July 2, 2010.
8. Jon Mooallem, "Do-It-Yourself Genetic Engineering," *New York Times Magazine*, February 10, 2010.
9. David Baltimore interview by Sarah Lippencott, Pasadena, California, October 13, November 19 and 25, 2009. Oral History Project, California Institute of Technology Archives. http://resolver.caltech.edu/CaltechOH:OH_Baltimore_D.
10. Shane Crotty, *Ahead of the Curve: David Baltimore's Life in Science* (Berkeley: University of California Press, 2003).
11. V. R. Racaniello and D. Baltimore, "Molecular Cloning of Poliovirus cDNA and Determination of the Complete Nucleotide Sequence of the Viral Genome," *Proceedings of the National Academy of Sciences* 78, no. 8 (1981), 4887–91.
12. Vincent Racaniello, "Thirty Years of Infectious Enthusiasm," www.virology.ws/2011/08/03/thirty-year-of-infectious-enthusiasm.

13. Tim Weiner, "Soviet Defector Warns of Biological Weapons," *New York Times*, February 5, 1998.
14. Deborah Josefson, "Scientists Manage to Manufacture Polio Virus," *BMJ* 325, no. 7356 (2002), 122.
15. Frank Gottron, "Synthetic Poliovirus: Bioterrorism and Science Policy Implications," Congressional Research Service Report for Congress, http://congressionalresearch.com/RS21369/document.php?study=Synthetic+Poliovirus+Bioterrorism+and+Science+Policy+Implications.
16. Jonathan B. Tucker, "The Smallpox Destruction Debate: Could Grand Bargain Settle the Issue?" *Arms Control Today*, March 2009.
17. David Whitehouse, "First Synthetic Virus Created," BBC News, July 11, 2002, http://news.bbc.co.uk/2/hi/2122619.stm.
18. Sanger Institute, www. sanger.ac.uk/resources/.
19. Centers for Disease Control and Prevention, "Outbreak of Ebola Hemorrhagic Fever—Uganda, August 2000–January 2001," *Morbidity and Mortality Weekly Report* 50, no. 5 (2001), 73–77.
20. Alston Chase, "Harvard and the Making of the Unabomber," *Atlantic*, June 2000.
21. Charles Mann, *1491: New Revelations of the Americas Before Columbus* (New York: Knopf, 2005), 92.
22. Steven Block, "The Growing Threat of Biological Weapons," *American Scientist*, January–February 2001.
23. Steve Sternberg, "Henderson Led WHO's Effort to Rid the World of Smallpox," *USA Today*, June 30, 2009.
24. Richard Preston, "The Bioweaponeers," *New Yorker*, March 9, 1998.
25. Tara O'Toole et al., "Shining Light on 'Dark Winter,'" *Clinical Infectious Diseases* 34, no. 7 (2002), 972–83.
26. Jonathan B. Tucker, *Scourge: The Once and Future Threat of Smallpox* (New York: Grove Press, 2001).
27. Richard Knox, "Manufacturing Problems with Swine Flu Vaccine," Shots, July 13, 2009, http://www.npr.org/blogs/health/2009/07/manufacturing_problems_with_sw.html.
28. Steve Sternberg, "U.S. Government Stockpiles New, Safer Smallpox Vaccine," *USA Today*, May 25, 2010.
29. "Feral Animals in Austalia," Australian Government Department of Stability, Environment, Water, Population and Communities, www.environment.gov.au/biodiversity/invasive/ferals/index.html.
30. Jon Cohen, "Designer Bugs," *Atlantic*, July–August 2002.
31. Nanhal Chen et al., "The Genomic Sequence of Ectromelia Virus, the Causative Agent of Mousepox," *Virology* 317, no. 1 (2003), 165–86.
32. "Venter Institute Scientists Create First Synthetic Bacterial Genome," J. Craig Venter Institute press release, January 24, 2008, www.jcvi.org/cms/research/projects/synthetic-bacterial-genome/.
33. Jerry Gordon, "Syria's Bio-Warfare Threat: An Interview with Dr. Jill Dekker," *New English Review*, December 2007.
34. ScienceWatch interview with Peter Palese, November 2009, http://sciencewatch.com/ana/st/h1n1/09novSTh1n1Pale/.

35. National Institute of Allergy and Infectious Diseases, "Reverse Genetics: Building Flu Vaccines Piece by Piece," http://www.niaid.nih.gov/topics/Flu/Research/vaccine Research/Pages/ReverseGenetics.aspx.
36. Steven Block interview by author.
37. Melinda Wenner, "Regaining Lost Luster," *Scientific American*, January 2008, and Sheryl Gay Stolberg, "A Death Puts Gene Therapy Under Increasing Scrutiny," *New York Times*, November 4, 1999.
38. Laura A. Shackleton, "High Rate of Viral Evolution Associated with the Emergence of Carnivore Parvovirus," *Proceedings of the National Academy of Sciences* 102, no. 2 (2004), 379–84.
39. David Vaux, "ABT-737, Proving to Be a Great Tool Even Before It Is Proven in the Clinic," *Cell Death and Differentiation* 15 (2008), 807–08.
40. Ira Flatow, "Genetic Engineering Conference Kicks Off at MIT," *Talk of the Nation*, November 7, 2008, http://www.npr.org/templates/story/story.php?storyId=96747191.

CHAPTER 6

1. Tara Adrienne Estlin, "Using Multi-Strategy Learning to Improve Planning Efficiency and Quality," Ph.D. diss., University of Texas at Austin, 1998.
2. "New Scheduling Algorithms Help Satellites Plan Their Own Activities," NASA Jet Propulsion Laboratory press release, April 18, 2010, http://scienceandtechnology.jpl .nasa.gov/newsandevents/newsdetails/?NewsID=886.
3. Riva Richmond, "Malicious Software Program Attacks Industry," *New York Times*, September 25, 2010.
4. Steven Cherry, "How Stuxnet is Rewriting the Cyberterrorism Playbook," This Week in Technology, October 13, 2010, http://spectrum.ieee.org/podcast/telecom/security/ how-stuxnet-is-rewriting-the-cyberterrorism-playbook.
5. On July 7, 2009, Reuters published an article, "Wary of Naked Force, Israel's Eye Cyberwar in Iran," by reporter Dan Williams speculating on whether Israel might launch a cyber attack on Iran's uranium enrichment program—a year before Stuxnet had surfaced. The article quoted Scott Borg extensively. When I asked Langner if he remembers having read this article, he said that he hadn't.
6. W32.Stuxnet Dossier, version 1.4, February 2011, www.symantec.com.
7. John Markoff, "A Silent Attack, but not a Subtle One," *New York Times*, September 26, 2010.
8. www.nei.org/newsnadevents/conferencesandmeetings/nnsc/.
9. "Aftermarket: Why Good Employees Violate Procedures—Is it Inevitable," Aviation Today, February 1, 2007, www.aviationtoday.com/am/categories/maintenance/ Aftermarket-Why-Good-Employees-Violate-Procedures-Is-it-Inevitable_8188.html. "A 2005 Internet survey of operational aviation professionals provided an interesting and troubling aspect of the loss equation. The survey indicated 63 percent of those professionals estimated intentional non-compliance events were occurring on a frequent to occasional basis."
10. William J. Broad and David E. Sanger, "Worm in Iran Was Perfect for Sabotaging Centrifuges," *New York Times*, November 19, 2010.

11. Winn Schwartau, *Terminal Compromise: Computer Terrorism in a Networked Society* (Seminole, FL: Inter-pact Press, 1991).
12. Charles G. Billo and Welton Chang, "Cyber Warfare: An Analysis of the Means and Motivations of Selected Nation States," Institute for Security Technology Studies at Dartmouth College report, November 2004, www.ists.dartmouth.edu/ISTS/library/briefings/ist0704.pdf.
13. James Glanz et al., "90 Seconds That Left Tens of Millions of People in the Dark," *New York Times*, August 26, 2003.
14. The video of the exploding generator, which originally appeared on CNN, is available on YouTube.com. Jeanne Meserve, "Staged cyber attack reveals vulnerability in power grid," CNN.com, September 26, 2007, http://articles.cnn.com/2007-09-26/us/power.at.risk_1_generator-cyber-attack-electric-infrastructure?_s=PM:US.
15. John Villasenor, "The Hacker in Your Hardware," *Scientific American*, August 2010.
16. Electricity Consumers Resource Council, "The Economic Impacts of the August 2003 Blackout," February 9, 2004, http://www.elcon.org/Documents/EconomicImpactsOfAugust2003Blackout.pdf.
17. Institution of Electrical and Electronic Engineers, "Nuclear Power: In the Wake of Three-Mile Island," *IEEE Spectrum* 21, no. 4 (1984).
18. According to Microsoft Security Intelligence Report, vol. 11, 1.8 million PCs were infected as of the second quarter of 2011.
19. Scott Borg, "A Cybersecurity Nightmare," *Scientific American*, November 2011.
20. John Markoff, "Scientists Worry that Machines May Outsmart Man," *New York Times*, July 26, 2009.
21. Michael Anderson and Susan Leigh Anderson, "Robot Be Good: A Call for Ethical Autonomous Machines," *Scientific American*, October 2010.
22. Jason Palmer, "Autonomous Tech 'Requires Debate,'" BBC News, August 19, 2009, http://news.bbc.co.uk/2/hi/technology/8210477.stm.
23. Michel M. Maharbiz and Hirotaka Sato, "Cyborg Beetles: Merging of Machine and Insect to Create Flying Robots," *Scientific American*, December 2010.
24. John Villasenor, "Here Come the Drones," *Scientific American*, January 2012.
25. Nadrian C. Seeman, "Nanotechnology and the Double Helix," *Scientific American*, June 2004.

INGENUITY

1. I'm describing the pieces in Freeman Dyson's *Infinite in All Directions* (New York: HarperCollins, 1988) and *Disturbing the Universe* (London: Boosey and Hawkes, 1979). I purchased the latter at a bookstore.
2. Dyson, *Infinite in All Directions*.
3. Freeman J. Dyson, "Warm Blooded Plants and Freeze Dried Fish," *Atlantic*, November 1997.
4. Nicholas Dawidoff, "The Civil Heretic," *New York Times Magazine*, March 2, 2009. Michael Lemonick followed up with an interview with Dyson, "Freeman Dyson Takes on the Climate Establishment," June 4, 2009, http://e360.yale.edu/content/feature.msp?id=2151.

5. Editors, "Fear and Its Consequences: Why States Should Get Tough with Vaccinations," *Scientific American*, February 2011.
6. Michael Specter, "A Life of Its Own: Where Will Synthetic Biology Lead Us?" *New Yorker*, September 28, 2009.
7. James Fallows, "Dirty Coal, Clean Future," *Atlantic*, December 2010.
8. Michael Lemonick, "A New Effort to Capture Coal's Dirty Breath & Bury It Underground," *Discover*, June 2010.
9. David R. Huggins and John P. Reganold, "No-Till: The Quiet Revolution," *Scientific American*, July 2008.
10. Justin Gillis and Celia W. Dugger, "U.N. Forecasts 10.1 Billion People by Century's End," *New York Times*, May 3, 2011.
11. Jeff Bartholet, "Inside the Meat Lab" *Scientific American*, June 2011.
12. Paul R. and Anne H. Ehrlich, *One With Nineveh: Politics, Consumption, and the Human Future* (Washington, D.C.: Island Press, 2004).

Index

acid rain, 34–35, 40
adaptability, 31–32
Africa
  agriculture and politics, 180
  Ebola virus, 8, 104
  population, 179–180
  West African monsoon, 68–70
  wheat failures, 92–93
agriculture
  artificial meat, 181
  big, centralized farms, 178
  and flu viruses, 1–2, 9–13
  food supplies, vulnerability of, 90–94
  genetically modified crops, 180–181
  lack of genetic diversity in food crops, 91–92
  no-till, 179
  perennial crops, 178–179
  population and, 179–180
  small-scale farms, 179
air circulation patterns, 75
Alaska, Sanak Island ecosystem, 84–89
algae, 51–55, 77–78
Alley, Richard, 64
Alvarez, Luis and Walter, 33
Amazon rain forests, 74–75
anaerobic bacteria, 42–43, 77–78
ant (*Myrmica sabuleti*), 29–30
Antarctic ice sheet, 72–74
anthrax, 102–103
anti-technology movement, 175–176
anti-vaccine movement, 175
apoptosis, 117–120
artificial intelligence and robotics, 156–165
  autonomous cars and airplanes, 163
  autonomous decision making, 126–130
  dangers of autonomy, 163–164
  drone technology, 161–162
  face recognition, 163
  IBM's Watson computer on *Jeopardy!*, 159
  machines taking over the world, 156–158
  Mars rovers, 128–129
  nanomachines, 164–165
  natural language abilities and impersonating people, 159–161
  privacy issues, 158–159, 162

artificial meat, 181
Association for the Advancement of Artificial Intelligence, 156–157
asteroid watch, 33–34
atmospheric circulation patterns, 75
Aum Shinrikyo, 97
Aurora simulated cyber attack, 145
Australia
  mousepox experiments, 109–112
  weather patterns, 78–81
autonomous programming, 126–130
avian influenza (bird flu), 9–17
  Hong Kong outbreak, 14–15
  mutation and the jump to humans, 1–2, 183
  Pennsylvania outbreak (1983), 10–12, 16–17
  virus swapping, 15
  waterfowl and human influenza, 9

bacteria
  anthrax, 102–103
  Great Oxidation Event, 42–43
Baltimore, David, 100–102
banking system, 154
bats, 90, 182
BCL-2, 118
beetles, computer chips in, 162
Besser, Richard, 18–19, 20–21
biobricks, 122
Biological Weapons Convention (1972), 101
bioweapons
  anthrax, 102–103
  apoptosis and, 117–120
  Baltimore and RNA viruses, 100–102
  Block on increasing potential for, 96–100
  escape viruses, 116–117
  gene therapy, 115
  Pentagon and the JASONS and, 95–96
  smallpox (*see* smallpox)
  stealth viruses, 116
  Wimmer's synthesis demonstration, 102
bioweapons labs, 113
bird flu. *See* avian influenza
Black Death, 25–26
blackouts
  eastern U.S. (2003), 144–145, 149
  societal and economic effects of, 149, 152–154
Blackwatch (computer program), 115
Black Watch (select agent program), 112
Block, Steven, 96–100, 113, 114–115, 119–120, 183–184
bluefin tuna, 89–90
boreal forests, 75–76
Borg, Scott, 130–132, 140, 144–45, 152, 153, 154–155
Borlaug, Norman, 91, 178
botnets, 132
brachiopods, 37
Brower, Kenneth, 169
Brown, Lester, 176
Bryozoans, 36
Byford, Ann, 122–125

Cambrian explosion, 41
Canada
  boreal forests, 75–76
  cod fishing, 89
canine parvovirus, 116, 117
carbon dioxide, 40, 44, 56–58
carbon emissions, 171–172, 176–177
cat distemper virus, 116–117
CED-3, 118
Centers for Disease Control (CDC), 18, 107–108, 112
Chicxulub meteorite impact, 33–36
China
  avian influenza in, 1–2, 9, 13–15
  coal plants, 177
  global effects of drought in, 81–83
  and grain shortages/prices, 93
Claas, Eric, 14
climate change, 50–83, 186
  Australian weather patterns, 78–81
  China, drought and water shortages, 81–83
  cutting emissions, 171–172, 176–177
  Dyson on, 169–170
  Earth's tilt and orbit and, 59–60
  emotional debate on, 169–170
  extinction and, 42–43, 186
  feedback loops, 60–61
  geoengineering/climate engineering, 170–174
  Hansen and runaway global warming, 56–58
  ice and, 60
  Lenton's tipping points, 64–78
    Amazon rain forest, 74–75
    Antarctic ice sheet, 72–74
    Canadian boreal forest, 75–76
    connections between tipping points, 76–77
    defining tipping points, 65–66
    El Niño–Southern Oscillation, 76
    Great Oxidation Event, 77–78
    Greenland's glaciers, 70–72
    Indian summer monsoon, 67–68
    north pole sea ice, 70
    West African monsoon, 68–70
  periods of rapid warming and cooling, 61–64
  Scheffer and pond eutrophication as microcosm, 50–56
  sequestering carbon, 177
climate cycles, 59–60
climate engineering/geoengineering, 170–174
cloning, 101
Clovis people, 46, 191n.18
coal, 40, 176–177
cod, 89
colony collapse disorder, 93, 182
computers. *See also* artificial intelligence and robotics; malware
  autonomous, 126–130
  Internet vulnerability, 3, 130–132, 184–185

computers (*continued*)
machines taking over the world, 142–143, 156–157
Conficker A, 155–156
Conkling, Philip (*The Fate of Greenland*), 60–61
corn, 91–92
crabs, 148
Crutzen, Paul, 171
cyanobacteria (pond scum), 42–43
cyber attacks, 131–132
cyber attack simulations, 143–144, 145

Daily, Gretchen, 182
Dark Winter, 107–108
Darwin, Charles, 31, 41
designer diseases, 117–120
dinosaurs, 32–36
distemper virus, 116–117
DNA, 95–96, 120, 121, 164–165
DNA sequences, 115
DNA viruses, 113–114
drone technology, 161–162
droughts, 68–70, 78–83
dual-use technology, 98–99, 160–161
Dunne, Jennifer, 84–89
dynamical systems, 52–56
Dyson, Freeman, 167–170

E. coli, 122–125
Eaarth (McKibben), 175
Earth Observing-I satellite, 130
Earth's tilt and orbit and, 59–60
Ebola, 8, 104
echinoderms, 36
economy
cyber attacks and, 143–144
effects of large-scale blackouts, 149, 151, 152–155
Internet vulnerability and, 130–132
stock market flash crash (May 6, 2010), 185
ecosystems
and food crops, 90–94
loss of predators and, 90
overfishing, 89–90
Sanak Island food web, 84–89
signs of decline in, 89–90
Ediacaran episode, 41
electricity grid, 144–154
destruction and replacement of generators, 149–152
eastern U.S. blackout (2003), 145
possible infection scenario, 147–149
simulated cyber attack (Aurora), 145
smart grid technology, 146
snowstorm and power outage (October, 2011), 5–6
societal and economic effects of blackouts, 149, 152–154
vulnerability from computer chips manufactured abroad, 146–147
El Niño Southern Oscillation, 64, 76, 81
environmentalism, 169–170, 176
Erik the Red, 62
Erwin, Doug, 37, 38–39, 43, 44, 48, 190n.14
escape viruses, 116–117

Estlin, Tara, 126–130
eutrophication, 51–55
extinction, 28–49
 adaptability and, 31–32
 and climate change, 185
 Ediacaran episode, 41
 explosion of life forms following extinction events, 35–36, 41–42
 Great Oxidation Event, 42–43, 77–78
 Holocene extinction, 43–49
  beginning of human impact, 45
  evidence for, 43–44
  Hawaiian extinctions, 48
  numbers of extinct species, 49
  Pleistocene mammals, 45–48, 191n.18
 K-T extinction event, 32–36
 Large Blue Butterfly, 28–30
 mass extinction events, defined, 27, 32
 Permian extinction event, 36–41, 190n.14
 species interdependence and, 30–31

face recognition, 163
Fate of Greenland, The (Conkling), 60–61
Fate of the Earth (Schell), 34
feedback loops, 60–61
Fichtner, Gerald, 10–11
fish, 54, 89–90
flash crash (May 6, 2010), 185
flips. *See* tipping points
flu. *See* influenza
follicular lymphoma, 118
food chains, 85
food supplies, 90–94. *See also* agriculture
food webs, 85–87
foraminifera, 36, 37
forest fires, 76
fungus
 Sporormiella, 47
 wheat rust, 92–93

Gaia hypothesis, 59, 64
Galápagos finches, 31
gas and oil industry, 154
Gaston Day School iGEM team, 120–125
generators, 145, 149–152
GeneSwitch, 116
gene therapy, 115
genetically modified crops, 180–181
genetic devices, 99–100
genetic engineering, 97, 182
genetic sequencing, 95–96, 100–105, 112–114, 120–121
geoengineering/climate engineering, 170–174
Gill, Jacquelyn, 47, 191n.18
glaciers and ice sheets, 60–61, 62, 72–74
Gleick, Peter, 79, 83
global warming. *See* climate change
golden rice, 180–181
Grand Coulee Dam, 150–151
Great Drought of the Aughts, 79
Great Oxidation Event, 42–43, 77–78

greenhouse effect, 40, 56–58, 190n.14
Greenland
early settlers of, 62–64
ice sheet, 60–61, 62, 70–72
Green Revolution, 91, 178
ground sloth, 46
Guan Yi, 14–15

Hansen, James, 56–58, 169, 176, 177
Hawaii, extinctions, 48–49
Henderson, D. H., 105–106
Hillis, Danny, 184, 185
H1N1. *See* swine (H_1N_1) influenza pandemic
H5N1 influenza, 14–15
Holocene extinction, 43–49
beginning of human impact, 45
evidence for, 43–44
Hawaiian extinctions, 48–49
numbers of extinct species, 49
Pleistocene mammals, 45–48, 191n.18
Homeland Security, U.S. Department of, 145–46
honeybees, colony collapse disorder, 93, 182
Hong Kong influenza outbreak, 14–15
Horvitz, Eric, 163
human genome, 95–96, 120
hunter-gatherer societies, 84–89
Hussein, Sheran, 120–125

IBM's Watson computer on Jeopardy!, 159
ice sheets and glaciers, 60–61, 62, 72–74
iGEM contest, 99–100, 120–125
Indian summer monsoon, 67–68
Infinite in All Directions (Dyson), 167–170
influenza viruses, 20–21. *See also* avian influenza (bird flu)
as bioweapons, 97–98
characteristics of, 13–14
Hong Kong (H5N1) outbreak, 14–15
infection sequence, 15–16
lethality and transmissibility traits, pandemic and, 2, 16, 97–98, 117
mutability of, 16–17
reverse engineering, 114
swine (H1N1) influenza pandemic
1918, 19, 26
2009, 17–21
2009, alternate scenario, 21–25
vaccines, 12–13, 23
interdependency, 30–31, 52–56. *See also* ecosystems
Internet, 3, 130–132, 184–185
Invitrogen, 116
iPhones, 158
Iran, uranium centrifuges. *See* Stuxnet malware
"Is There Still Time to Avoid Dangerous Anthropogenic Interference with Global Climate?" (Hansen), 57

Jackson, Ron, 109–112
jaguars, 181–182
JASONS, 95–97
Jennings, Ken, 159

Jeopardy!, 159
Jet Propulsion Laboratory, 128–129
Johnson, Scott, 69
Journal of Virology, 112
Joy, Bill, 184–185

Kaczynski, Ted, 104–105
Kahn, Bob, 184–185
Kalengyere Research Station (Uganda), 92
Kawaoka, Yoshi, 7–10, 12, 16–17
Keasling, Jay, 175
Keeling, Charles David, 56–57
Keith, David, 170–174
Kelly, John, 26
KevaCon International, 119
Khan, A. Q., 140
Kida, Hiroshi, 7
K-T extinction event, 32–36
Kurzweil, Ray, 164

"Lake as a Microcosm, The" (Forbes), 51–52
lakes and ponds as microcosms, 51–56
Langner, Ralph, 133–136, 141
La Niña, 76
Large Blue Butterfly, 28–30
Layton, Marci, 19–20
Lederberg, Joshua, 9
Lenton, Tim, 64–77
  Amazon rain forest tipping point, 74–75
  Antarctic ice sheet tipping point, 72–74
  Canadian boreal forest tipping point, 75–76
  connections between tipping points, 76–77
  defining tipping points, 65–66
  El Niño–Southern Oscillation tipping point, 76
  on Great Oxidation Event, 77–78
  Greenland's glaciers tipping point, 70–72
  Indian summer monsoon tipping point, 67–68
  north pole sea ice tipping point, 70
  West African monsoon tipping point, 68–70
LiveWire cyber attack simulation, 143–144
London, Thames Barrier, 73–74
Lovelock, James, 59, 64
Lyons, Timothy, 43

malware
  and autonomous computers, 164
  banking system and, 154
  Conficker A, 155–156
  cyber attack simulation (LiveWire), 143–144
  difficulties of preventing, 155–156, 184
  electricity grid and, 144–154
    destruction and replacement of generators, 149–152
    possible infection scenario, 147–149
    simulated cyber attack (Aurora), 145
    smart grid technology, 146
    societal and economic effects of blackouts, 149, 152–154

malware (continued)
vulnerability from computer chips manufactured abroad, 146–147
gas and oil industry and, 154
Internet vulnerability, 130–132
Stuxnet malware, 132–143
damage caused by, 140–141
discovery of, 132–134
eradication of, 141–142
importance of, 136
method of infection, 136–139
purpose of, 134–136
telecommunications and, 154
mammals, rise of, 35–36
mammoths, 47–48, 191n.19
Mars rovers, 128–129
Massachusetts Institute of Technology (MIT) iGEM contest, 99–100, 120–125
mass extinction events, 27, 32
Mass Fatality Plan (U.S. Department of Health and Human Services), 23–24
McKibben, Bill, 175
meat, artificial, 181
MER Opportunity Rover, 129
meteorite impacts, 35–36
methane, 40, 66
microbial genomes, 103–104
Milankovitch cycles, 60
"Minimal Models as Useful Tools for Ecologists" (Scheffer), 55
Mitchell, Tom, 157–158
MIT iGEM contest, 99–100, 120–125
monsoons, 67–70
Mount Pinatubo eruption, 35, 40, 171
mousepox, 109–112
Murray River (Australia), 79

nanomachines, 164–165
NASA
asteroid watch, 33–34, 35
Earth Observing-I satellite, 130
Mars rovers, 128–129
natural language abilities, 159–161
Netherlands, rising sea levels and, 72–73
Neumann, Gabi, 114
New Jersey Sierra Club, 177
New Scientist, 112
Newsweek, 69, 102–103
Newsweek International, 1
New York Times Magazine, 169
nitrates, 122–123
north pole sea ice, 70
no-till agriculture, 179
Nowak, Rachel, 112
nuclear winter, 34

optimism, 4–5
oxygen, Great Oxidation Event, 42–43

Pacala, Stephen, 4
Palese, Peter, 114
Parkman, Breck, 190n.16
partial-order planning, 126–127
parvovirus, 116–117
passenger pigeon, 44
Penfield, Glen, 33
Pennsylvania avian influenza outbreak (1983), 10–12, 16–17

Pentagon, 95–96
perennial food crops, 178–179
permafrost, 66, 76
Permian mass extinction event, 36–41, 190n.14
Permian reef, 36–38
Phillips, John, 37–38
plasmids, 114, 123
Pleistocene mammals, 45–48, 191n.18
polio virus, 101
ponds and lakes as microcosms, 51–56
population, 3, 179–180, 182
power outages. *See* electricity grid
predators, loss of, 90
prey-switching behavior, 88, 89
Proceedings of the National Academies of Sciences, 170
PurGen, 177

rain forests, 74–75
Ramshaw, Ian, 109–112
reverse genetics, 114–115
reverse transcriptase, 101
RNA viruses, 100–102, 113. *See also* influenza viruses
robots. *See* artificial intelligence and robotics
rovers, 128–129

Sacculina parasite, 148
Sahara desert (Africa), 69–70
Sahel region (Africa), 68–70
Sanak Island ecosystem, 84–89
sarin nerve gas, 97
SARS, 1
Schafer, Sarah, 1
Scheffer, Marten, 50–56, 58
Schell, Jonathan, 34
Schuchat, Anne, 18
Schwartau, Winn, 143
Scientific American, 64
sea levels, 70–74
Seeman, Ned, 164–165
Select Agents and Toxins List, 112
short-faced bear, 46
Shortridge, Ken, 14
Shrag, Dan, 176–177
Siberian volcanoes (Permian era), 39–40, 190n.14
Siemens controller chips, 133–134
Singularity Is Near, The (Kurzweil), 164
smallpox virus, 104–115
  characteristics of, 105
  eradication efforts, 105–106
  mousepox experiments and, 109–112
  obstacles to use as bioweapon, 104–105, 112–115
  reverse genetics and, 114–115
  scenario of terrorists use of, 106–107
  simulated attack (Dark Winter), 107–108
  vaccines, 108–109, 111–112
small-scale agriculture, 179
smart grid technology, 146
snowball Earth, 57–58
snowstorm and power outage (October 2011), 5–6
Southeast Asia climate shifts, 64
Soviet Union, smallpox manufacturing, 106
species, numbers of, 31, 49

stealth viruses, 116
stem rust, 92–93
stock market flash crash (May 6, 2010), 185
Stuxnet malware, 132–143
  damage caused by, 140–141
  discovery of, 132–134
  eradication of, 141–142
  importance of, 136
  method of infection, 136–139
  purpose of, 134–136
sulfate aerosols, 171
sulfur, 170–171
swine (H1N1) influenza pandemic
  1918, 19, 26
  2009, 17–21
  2009, alternate scenario, 21–25
Symantec, 135–136
synthetic biology, 99, 120–125, 183–184

telecommunications, 154
Terminal Compromise (Schwartau), 143
Texas, Permian reef, 36–38
Thames Barrier, 73–74
Thrun, Sebastian, 163, 165
tipping points, 64–78
  Amazon rain forest, 74–75
  Antarctic ice sheet, 72–74
  Canadian boreal forest, 75–76
  connections between tipping points, 76–77
  defining, 65–66
  El Niño–Southern Oscillation, 76
  Great Oxidation Event, 77–78
  Indian summer monsoon, 67–68
  lake eutrophication, 52–54, 58
  north pole sea ice, 70
  runaway global warming, 56–58
  West African monsoon, 68–70
Toenniessen, Gary, 180
tuna, 89–90
turbidity, 51–55

uranium centrifuges, 134–141
Utah, Green River fossils, 37–38
Utrecht, University of, 51, 55

vaccines, 23, 108–109, 111–112, 175
value destruction, 131–132, 152–153
Vaux, David, 118
Venter, Craig, 96, 112, 175–176
Venus effect, 56, 58, 65–66
Vikings, 62–64
VirusBlokAda (Belarus), 133
viruses
  apoptosis and, 117–120
  as bioweapons, 96–102
  Ebola, 8, 104
  escape viruses, 116–117
  genetic sequencing of, 103–104
  influenza (*see* avian influenza; influenza viruses)
  RNA viruses, 100–102, 113
  smallpox (*see* smallpox)
  stealth viruses, 116
viruses, computer. *See* malware
Virus Research, 104

Webster, Robert, 8–10, 11–12, 13, 17

West African monsoon, 68–70
wheat, 91–93
white box setup, 124
White-Nose Syndrome, 90
wildfires, 80. *See also* forest fires
wildlife, 181–182
Wilson, E. O., 48
Wimmer, Eckhard, 102
Wolderwijd, Lake, 54–55
wolves, 90
World Health Organization (WHO), 17, 105–106
worst-case scenarios
  electricity grid, 147–149
  swine (H1N1) influenza pandemic scenario, 21–25
  terrorists use of smallpox virus, 106–107
  value of, 4

Yellowstone National Park
  wolves, 90

zooplankton, 53–54

A Note on the Author

Fred Guterl is the executive editor of *Scientific American*. For ten years he edited *Newsweek International*, most recently as deputy editor. Guterl holds a bachelor of science degree in electrical engineering from the University of Rochester and has taught science writing at Princeton University. He lives near New York City with his wife and two children.

Guterl, Fred.

The fate of the species.